# Lecture Notes in Control and Information Sciences

Edited by M. Thoma

For information about Vols. 1–42 please contact your bookseller or Springer-Verlag.

# Lecture Notes in Control and Information Sciences

Edited by M. Thoma and A. Wyner

101

P. E. Crouch
A. J. van der Schaft

Variational and Hamiltonian
Control Systems

Springer-Verlag Berlin Heidelberg GmbH

**Authors**
Dr. P. E. Crouch
Dept. of Electrical and Computer Engineering
Arizona State University
Tempe, AZ 85287
USA

Dr. A. J. van der Schaft
Dept. of Applied Mathematics
University of Twente
P. O. Box 217
7500 AE Enschede
The Netherlands

Library of Congress Cataloging in Publication Data
Crouch, P. E.
Variational and Hamiltonian control systems.
(Lecture notes in control and information sciences; 101)
Bibliography: p.
1. Control theory.
2. Calculus of variations.
3. Hamiltonian systems.
I. Schaft, A. J. van der.
II. Title
III. Series.
QA402.3.C74    1987    629.8'312    87-26421
ISBN 978-3-540-18372-3      ISBN 978-3-540-47929-1 (eBook)
DOI 10.1007/978-3-540-47929-1

© Springer-Verlag Berlin Heidelberg 1987
Originally published by Springer-Verlag Berlin Heidelberg New York in 1987.
Offsetprinting: Mercedes-Druck, Berlin

2161/3020-543210

# PREFACE

This monograph grew out of a combined effort to prove a conjecture, formulated by the second author, concerning the characterization of Hamiltonian control systems in terms of their variational input-output behaviour. This conjecture was motivated by the Hamiltonian Realization Problem as well as by the Inverse Problem in Classical Mechanics. In the course of proving a slightly modified version of this conjecture we developed some concepts, whose interest seems not to be confined to Hamiltonian control systems. In particular the concepts of the prolonged system and the Hamiltonian extension, based on considering the variational and the adjoint variational control systems to any nonlinear system, are, as we believe, of independent interest to control theory.

The main concepts and results of this monograph are contained in chapters (1) to (6). In chapter (0) we give a brief introduction to Hamiltonian control systems, with particular emphasis on the relations between physical and control theoretic notions. Indeed, the study of Hamiltonian control systems is one of the places where (theoretical) physics and system and control theory meet. We conclude the monograph with chapter (7) discussing some possible extensions to the theory presented, as well as some open problems.


Tempe,
Enschede, June 1987

# CONTENTS

# 0. INTRODUCTION

Of central importance in the modelling of physical systems are the classical Euler-Lagrange or Hamiltonian equations. These equations describe the dynamics of a very large class of conservative physical systems, including mechanical and electromagnetic systems, and lie at the heart of the theoretical framework of most physics. Although the conservation of energy is usually an idealization, in many cases the neglection of dissipation of energy (friction, damping) forms a natural starting point.

Let us consider, for example, a conservative mechanical system with n degrees of freedom, locally represented by n (generalized) configuration variables $q_1, \ldots, q_n$. The Euler-Lagrange equations are the following well-known set of second-order differential equations

$$(0.1) \qquad \frac{d}{dt} \left( \frac{\partial L}{\partial \dot{q}_i} \right) - \frac{\partial L}{\partial q_i} = F_i \qquad\qquad i = 1, \ldots, n$$

where $L(q, \dot{q}) = L(q_1, \ldots, q_n, \dot{q}_1, \ldots, \dot{q}_n)$ is the Lagrangian of the system. In most mechanical systems the Lagrangian is the difference of a kinetic energy $T(q, \dot{q})$ and a potential energy $V(q)$

$$(0.2) \qquad L(q, \dot{q}) = T(q, \dot{q}) - V(q)$$

where $T(q, \dot{q})$ is quadratic in the generalized velocities

$$(0.3) \qquad T(q, \dot{q}) = \frac{1}{2} \dot{q}^T M(q) \dot{q}$$

for some positive-definite matrix $M(q)$. In this case the Euler-Lagrange equations specialize to

$$(0.4) \qquad \frac{d}{dt} \left( \frac{\partial T}{\partial \dot{q}_i} \right) - \frac{\partial T}{\partial q_i} = - \frac{\partial V}{\partial q_i} + F_i, \qquad i = 1, \ldots, n$$

and the terms $- \frac{\partial V}{\partial q_i}$ represent the <u>internal</u> conservative (i.e. derivable from a potential) forces in the system. Finally the vector $F = (F_1, \ldots, F_n)$ denotes the (generalized) <u>external</u> forces acting on the system while in configuration $(q_1, \ldots, q_n)$.

In the (mathematical) physics literature the external forces usually are seen as

given functions of time. Consequently, the external forces are often split into
two parts: one "maximal" component which is derivable from a potential function,
and so can be added to the internal forces, and remaining non-conservative forces.
Alternatively, in stochastic mathematical physics the external forces are modelled
as stochastic variables [B1]. In systems and control theory the approach, however,
is quite different. Usually (some of) the external forces will be interpreted as
control or input variables; i.e., "arbitrary" functions of time. Instead of considering the influence of the environment on the system as given, one is primarily
interested in the way the system will react to different external forces. Of
course this is intimately related to the fact that in control theory one wishes to
prescribe the behavior of the system, instead of only describing it [B1].

In general not all degrees of freedom are directly accessible to control action,
resulting in Euler-Lagrange equations of the form

$$(0.5) \qquad \frac{d}{dt}\left(\frac{\partial L}{\partial \dot{q}_i}\right) - \frac{\partial L}{\partial q_i} = \begin{cases} u_i & i = 1,\ldots,m \\ 0 & i = m+1,\ldots,n \end{cases}$$

where now $u = (u_1,\ldots,u_m)$ are the controls or inputs (i.e., "arbitrary" functions
of time). We call (0.5) a Lagrangian control system.

As is well-known, the Hamiltonian equations of motion are obtained from the Euler-
Lagrange equations (0.1) by defining the generalized momenta

$$(0.6) \qquad p_i = \frac{\partial L}{\partial \dot{q}_i} \qquad\qquad i = 1,\ldots,n$$

In most cases the transformation from $(\dot{q}_1,\ldots,\dot{q}_n)$ to $(p_1,\ldots,p_n)$ is a (local)
diffeomorphism, allowing us to transform the Lagrangian $L(q,\dot{q})$ into the Hamiltonian $H(q,p) = \sum_{i=1}^{n} p_i \dot{q}_i - L(q,\dot{q})$ (the Legendre transformation), and the (second-
order) Euler-Lagrange equations into the set of first-order differential equations

$$(0.7) \qquad \begin{aligned} \dot{q}_i &= \frac{\partial H}{\partial p_i} \\ \dot{p}_i &= -\frac{\partial H}{\partial q_i} + F_i \end{aligned} \qquad\qquad i = 1,\ldots,n$$

which are called the Hamiltonian equations of motion. In case the Lagrangian is
given as in (0.2)-(0.3) the Hamiltonian becomes

(0.8) $\qquad H(q,p) = \frac{1}{2} p^T M^{-1}(q)p + V(q)$

and so denotes the (internal) energy. Finally the Lagrangian control system (0.5) results in the Hamiltonian control system

(0.9)
$$\dot{q}_i = \frac{\partial H}{\partial p_i} \qquad\qquad i = 1,\ldots,n$$

$$\dot{p}_i = -\frac{\partial H}{\partial q_i} + \begin{cases} u_i & i = 1,\ldots,m \\ 0 & i = m+1,\ldots,n \end{cases}$$

For (0.5) as well as (0.9) we have assumed for simplicity that the inputs $u_i$, $i = 1,\ldots,m$, are directly coupled to the first m degrees of freedom. Of course this particular form is not invariant under a nonsingular change of configuration coordinates

(0.10) $\qquad q_i = \phi_i(\bar{q}_1,\ldots,\bar{q}_n) \qquad\qquad i = 1,\ldots,n$

with the Jacobian $D\phi(\bar{q})$ everywhere non-singular. In fact, as can be easily checked, under such a coordinate transformation the Lagrangian control system transforms into

(0.11) $\qquad \dfrac{d}{dt}\left[\dfrac{\partial \bar{L}}{\partial \dot{\bar{q}}_i}\right] - \dfrac{\partial \bar{L}}{\partial \bar{q}_i} = \displaystyle\sum_{j=1}^{m} u_j \dfrac{\partial \phi_j}{\partial \bar{q}_i} \qquad i = 1,\ldots,n$

with $\bar{L}(\bar{q},\dot{\bar{q}}) = L(q,\dot{q})$, while the Hamiltonian control system (0.9) becomes

(0.12)
$$\dot{\bar{q}}_i = \frac{\partial \bar{H}}{\partial \bar{p}_i}$$

$$\dot{\bar{p}}_i = -\frac{\partial \bar{H}}{\partial \bar{q}_i} + \sum_{j=1}^{m} u_j \frac{\partial \phi_j}{\partial \bar{q}_i} \qquad\qquad i = 1,\ldots,n$$

where $\bar{H}(\bar{q},\bar{p}) = H(q,p)$.

Let us from now on concentrate on Hamiltonian control systems. Notice that (0.12) suggests we enlarge the class of Hamiltonian control systems to systems of the form

$$(0.13) \qquad \dot{q}_i = \frac{\partial H_0}{\partial p_i}$$

$$\dot{p}_i = -\frac{\partial H_0}{\partial q_i} + \sum_{j=1}^{m} u_j \frac{\partial H_j}{\partial q_i} \qquad\qquad i = 1,\ldots,n$$

with $H_0(q,p)$ the internal Hamiltonian, and $H_j(q)$, $j = 1,\ldots,m$, arbitrary (smooth) functions. In particular, this form is clearly invariant under a change of configuration coordinates.

We shall even go a little bit further. Part of the power of Hamiltonian formalism is to regard the generalized momenta $p_i$ on the same footing as the generalized configuration coordinates $q_i$. Consequently, one does not only allow for transformations of the configuration coordinates $q_i$ (with $p_i = \frac{\partial L}{\partial \dot{q}_i}$ resulting in a transformation of the $p_i$), but one considers <u>all</u> coordinate transformations $(q,p) \to (\bar{q},\bar{p})$, which leave the Hamiltonian form of the equations invariant, i.e. the <u>canonical</u> transformations. Under such a general canonical transformation, the functions $H_1,\ldots,H_m$ become functions of q <u>and</u> p, and therefore we define a <u>general (affine) Hamiltonian control system</u> as

$$\dot{q}_i = \frac{\partial H_0}{\partial p_i} - \sum_{j=1}^{m} u_j \frac{\partial H_j}{\partial p_i}$$

$$(0.14) \qquad\qquad\qquad\qquad i = 1,\ldots,n$$

$$\dot{p}_i = -\frac{\partial H_0}{\partial q_i} + \sum_{j=1}^{n} u_j \frac{\partial H_j}{\partial q_i}$$

where the functions $H_0,H_1,\ldots,H_m$ are all arbitrary functions of q and p. (See Example 3 for a physical interpretation.)

Notice that a general Hamiltonian system (0.14) can also be regarded as a set of time-varying Hamiltonian differential equations governed by the time-varying Hamiltonian

$$(0.15) \qquad H_0(q,p) - \sum_{j=1}^{m} u_j(t)H_j(q,p)$$

This can be interpreted in the following abstract way (see also [Bu1,Bu2,V1]). The possibility of controlling the system with Hamiltonian (internal energy) $H_0(q,p)$ rests on the ability to exchange energy with the environment along some external channels. This can be regarded as the physical basis of contol. The exchangeable energy along the j-th channel is of the form $H_j(q,p)$, and $u_j$ denotes the strength of this energy exchange. Indeed, it is easily deduced that

(0.16) $\qquad \dfrac{dH_0}{dt} = \displaystyle\sum_{j=1}^{m} u_j(t)\, \dfrac{dH_j}{dt}$

In physics the Hamiltonians $H_j(q,p)$, $j = 1,\ldots,m$, are called <u>interaction</u> or <u>coupling</u> Hamiltonians.

Up to now we have not yet defined <u>outputs</u> y of a Hamiltonian control system. Of course nothing forbids us to consider as outputs arbitrary functions of the state $x = (q,p)$. However, there is a <u>natural</u> set of outputs associated to every Hamiltonian control system (0.14), namely the interaction Hamiltonians themselves:

(0.17) $\qquad y_j = H_j(q,p) \qquad\qquad\qquad j = 1,\ldots,m$

There are many good reasons for doing this. First of all with this choice of outputs we obtain from (0.16) the <u>energy balance</u>

(0.18) $\qquad \dfrac{dH_0}{dt} = \displaystyle\sum_{j=1}^{m} u_j \dot{y}_j\,.$

Hence the decrease or increase of the internal energy of the system is a function of the inputs and (the time-derivatives of) the outputs <u>only</u>. (Compare with the work of Willems on dissipativeness [W3].) For example in the simple case (0.9) where $y_j = q_j$, $j = 1,\ldots,m$, we have

(0.19) $\qquad \dfrac{dH_0}{dt} = \displaystyle\sum_{j=1}^{m} u_j \dot{q}_j$

and $\displaystyle\sum_{j=1}^{m} u_j \dot{q}_j$ equals the instantaneous external work performed on the system.

Secondly, with this particular choice of outputs we obtain the following symmetry or reciprocity between inputs and outputs. The external "forces" $u_1,\ldots,u_m$ influence the system via the external channels corresponding to the outputs $H_1,\ldots,H_m$, which are the "displacements" caused by these excitations along the same line of action. For example in case $H_j(q) = q_j$, $j = 1,\ldots,m$, the input $u_j$ equals the external force corresponding to the j-th degree of freedom $q_j$. Hence if $q_j$ is a Cartesian coordinate, then $u_j$ will be a translational force, while if $q_j$ is, say, an angular coordinate, then $u_j$ will be the corresponding external torque (see also Example 2). Notice also that in the original Euler-Lagrange or Hamiltonian equations (0.1) and (0.7) the vector $F = (F_1,\ldots,F_n)$ represents the external forces <u>as measured</u> in the configuration $(q_1,\ldots,q_n)$. Hence in order to define the external forces we need to know the configuration coordinates $(q_1,\ldots,q_n)$ which are (if we interpret $F_i$ as inputs) just the natural outputs of the system. Let us also remark that in order to define a general coupling Hamiltonian we need this to be a function of the observations made on the system, i.e. a function of the natural outputs. A third argument for choosing $y_j$ as in (0.16) has a more system-theoretic flavour. Consider for the Hamiltonian system (0.13) a state feedback law $u_j = \alpha_j(q,p) + v_j$, with $v_j$ the new inputs. When is the system after feedback again Hamiltonian? This is the case ([V1]) if and only if there exists a function S such that

$$(0.20) \qquad \alpha_j(q,p) = \frac{\partial S}{\partial y_j}(H_1(q,p),\ldots,H_m(q,p)) \qquad j = 1,\ldots,m$$

i.e., if and only if the feedback is output feedback with respect to the natural outputs and furthermore has the special form as in (0.20).

From a mathematical point of view this discussion can be summarized by noting that the space of inputs and natural outputs can be given the structure of a cotangent bundle T*Y, where Y is the output manifold with local coordinates $(y_1,\ldots,y_m)$, and where the coordinates of the fibers of this bundle are the inputs (or external forces) $u_i$. Concluding, we define a general (affine) Hamiltonian input-output system as a Hamiltonian control system (0.14) together with the natural outputs (0.17).

Finally let us notice the close similarity with the description of electrical networks with external ports. In this case each external port carries two "dual" variables, current and voltage, which are also needed for stating an energy (in-)equality. In our formalization of Hamiltonian systems the external channels involve the dual variables $u_j$ ("force") and $y_j = H_j$ ("displacement"). For more details, including a partial treatment of the theory of interconnections of Hamiltonian input-output systems we refer to [V1].

<u>Some examples</u>

1. Consider the following linear mass-spring system (without friction)

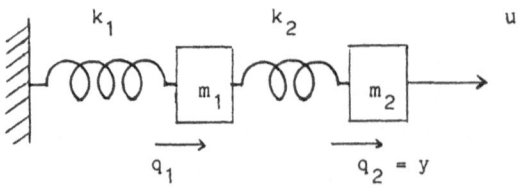

The Hamiltonian $H_0$ here is the sum of the kinetic and potential energies of both masses $m_1$ and $m_2$. If u is the external force on mass $m_2$ then the natural output y is the displacement $q_2$ of this second mass. The same holds for the first mass $m_1$.

2. Consider the following simple robot manipulator with two revolute joints

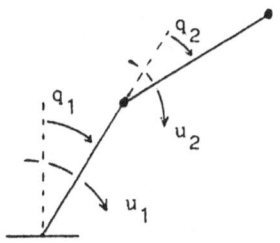

Both joints are equipped with an actuator delivering an external torque $u_i$, i = 1,2. Again $H_0$ is the total energy of the system. Notice that if we choose the outputs to be the Cartesian coordinates of the end-point, then the corresponding inputs would be the vertical and horizontal external forces on this point. In general, neglecting friction, robot manipulators (rigid or flexible) with an arbitrary number of degrees of freedom (revolute or prismatic) can be described as Hamiltonian input-output systems with u = $(u_1, \ldots, u_m)$ the forces as delivered by the actuators and y = $(y_1, \ldots, y_m)$ the corresponding configuration (joint) coordinates.

3. Consider the same system as in Example 1

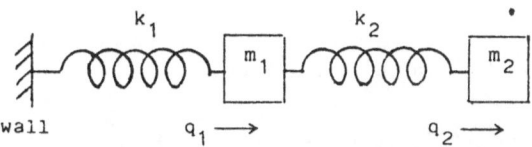

Now assume that the wall is movable, and that the velocity v of the wall can be controlled, i.e. u = v. The natural output y is now the momentum of mass $m_1$ (with respect to the inertial frame). Notice that $\dot{y}$ is the <u>reaction force</u>, and so $u\dot{y}$ is again the external power.

Up to now we have only considered Hamiltonian systems where the control variables $u_1, \ldots, u_m$ enter the equations in a linear (affine) way. Equivalently, we assumed all time-varying Hamiltonians to be of the form $H_0(q,p) - \sum_{j=1}^{m} u_j H_j(q,p)$. However, in its most general form the controlling action for a Hamiltonian system consists in arbitrary (not necessarily affine) dependence of the Hamiltonian on the controls. This means we have to consider general Hamiltonians $H(q,p,u)$ resulting in Hamiltonian control systems as originally proposed by Brockett [B1]

$$(0.21) \quad \dot{q}_i = \frac{\partial H}{\partial p_i} (q,p,u)$$

$$\dot{p}_i = - \frac{\partial H}{\partial q_i} (q,p,u) \qquad\qquad i = 1,\dots,n \qquad u = (u_1,\dots,u_m)$$

The natural outputs in this case are given as

$$(0.22) \quad y_j = c_j \frac{\partial H}{\partial u_j} (q,p,u), \qquad\qquad j = 1,\dots,m, \quad c_j = \pm 1.$$

System (0.21) with outputs (0.22) is called a general Hamiltonian input-output system, see [V1, V2]. Note that as a special case we recover static reciprocal systems

$$(0.23) \quad y_j = \frac{\partial H}{\partial u_j} (u) \qquad\qquad j = 1,\dots,m.$$

Examples (continued)

4. Consider k pointmasses $m_i$, with positions $q^i \in \mathbb{R}^3$, $i = 1,\dots,k$, in their own gravitational field corresponding to the potential

   $V(q^1,\dots,q^k) = \sum\limits_{i<j} \frac{m_i m_j}{\|q^i - q^j\|}$. Suppose that the controls are the positions of the first $\ell$ point masses $(\ell \leq k)$. The state then consists of the positions and momenta of the remaining $(k - \ell)$ point masses. The full Hamiltonian is given by

$$H(q^{\ell+1},\dots,q^k,p^{\ell+1},\dots,p^k,u^1,\dots,u^\ell) :=$$

$$V(u^1,\dots,u^\ell,q^{\ell+1},\dots,q^k) + \sum_{i=\ell+1}^{k} \frac{1}{2m_i} \|p^i\|^2$$

with $u^i = q^i \in \mathbb{R}^3$, $i = 1,\dots,\ell$. The resulting Hamiltonian control system is

$$\dot{q}_j^i = \frac{\partial H}{\partial p_j^i}$$

$$\dot{p}_j^i = - \frac{\partial H}{\partial q_j^i} \qquad\qquad i = \ell+1,\dots,k \qquad j = 1,2,3$$

with outputs

$$y_j^i = - \frac{\partial H}{\partial q_j^i} \qquad\qquad i = 1,\dots,\ell \qquad j = 1,2,3$$

which are the _forces_ experienced by the first $\ell$ point masses.

Apart from being strongly physically motivated, Hamiltonian control systems form a class of systems with very peculiar mathematical properties. The position of Hamiltonian control systems with regard to general (nonlinear) control systems should be compared to the position of Hamiltonian vectorfields with regard to general vectorfields. As is well-known there is much theory available on Hamiltonian vectorfields; theory which relies heavily on their particular structure. On the other hand Hamiltonian vectorfields form a _critical_ class of vectorfields, which is often not amenable to common approaches such as linearization techniques. Previous work on Hamiltonian control systems indicates the same tendency [V1, V3, M, V8, V9]. The fundamental notions of _symmetry_ and _conservation law_ can be directly extended to Hamiltonian control systems, resulting in a generalization of Noether's theorem. Preliminary work on stabilization and structural synthesis problems for Hamiltonian control systems shows that the Hamiltonian structure could be used in an advantageous way also in this context.

The main problem we wish to address in this monograph concerns the _characterization_ of Hamiltonian input-output systems, motivated by the Realization Problem in system theory, as well as by the classical Inverse Problem in mechanics. The Inverse Problem, which has a longstanding interest in (mathematical) physics, can be phrased in its simplest form as follows: How can we single out among all sets of second-order differential equations the Euler-Lagrange equations (with zero external forces)? Notice that the Inverse Problem has two aspects. First, what distinguishes the Euler-Lagrange equations from general second-order differential equations? Secondly, when can a set of second-order differential equations be rewritten as Euler-Lagrange equations, so that it "corresponds" to a conservative physical system? The practical importance of this second aspect, apart from its interest per se, lies in the fact that if we know a set of second-order differential equations to be Euler-Lagrange then all the powerful machinery developed for such equations may be applied to it. A more general motivation comes from theoretical physics, in particular quantization theory ([Sa1,Sa2]): Euler-Lagrange systems are at the basis of theoretical physics, and, in principle, can be quantized to quantum mechanical systems.

In the same vein, the question can be posed of determining when a nonlinear state space system with inputs and outputs

$$(0.23) \qquad \dot{x} = g_0(x) + \sum_{j=1}^{m} u_j g_j(x)$$

$$y_j = H_j(x) \qquad\qquad j = 1,\ldots,m$$

can be written as a _Hamiltonian_ input-output system. A second closely related question, which ties in with the Realization Problem in system theory, is the following. Given the input-output map of a system (or more generally its external behavior [W1, V1]); what conditions does this external behavior have to satisfy in order that it can be realized as the external behavior of a Hamiltonian input-output system? The problem of realizing an input-output map by a Hamiltonian input-output system is called the Hamiltonian Realization Problem.

Notice that both questions, as in the Inverse Problem, have two aspects. First, to specify the special properties possessed by Hamiltonian input-output systems, or by their input-output maps. Secondly, to determine which systems, respectively which input-output maps, can be realized by a Hamiltonian input-output system, respectively by the input-output map of a Hamiltonian system. Again this second aspect is of practical importance, since if we know a system to be equivalent to a Hamiltonian system, then we may use this extra information for control and synthesis purposes. Also Hamiltonian systems may be quantized to quantum mechanical control systems [Bu1,Bu2, V7, Ta]. Notice that the second aspect also contains a constructive element: Can a system given by a set of differential equations (0.23), or as an input-output map, be realized (in a physical sense) as the interconnection of some specific physical elements like (translational or rotational) springs, masses etc.? This element certainly is of some interest in itself. For instance, one may ask if a particular input-output behavior can be realized as the input-output behavior of a mechanical system like a robot manipulator. Let us remark that in linear electrical network theory there is an analogous question (realization of transfer matrices with electrical elements).

In this monograph we shall first deduce conditions under which a (minimal) nonlinear control system (0.24) with an equal number of inputs and outputs can be written as a Hamiltonian input-output system. The way these necesssary and sufficient conditions are found is of some interest in itself. We show how to each control system (0.24), with n states and m inputs and outputs, can be associated, in a natural way, two control systems with 2n states and 2m inputs and outputs. Roughly speaking, the first system is the original system together with all variational control systems, while the second system is the original system together with all the adjoint variational systems. These two canonical extensions are dual to each other in a precise sense. Now a (minimal) Hamiltonian input-output system is completely determined by the fact that these dual extensions are isomorphic systems. From (nonlinear) realization theory we know this to be equivalent to the two extensions having the same input-output behavior. The resulting conditions can be interpreted as generalizations of the conditions for solvability of the classical Inverse Problem, which, in their first form, were stated at the end of the last century (see for a discussion [Sa1]).

We then proceed to find conditions directly in terms of the variational input-output behaviour of the control system (0.23). In [V1, V2] already a conjecture was stated, which formed the original inspiration for this monograph. As a matter of fact, we shall be able to prove a slightly modified version of this conjecture. From a mathematical point of view the conditions of this conjecture can be expressed, roughly speaking, as follows. We can define a "symplectic form" on the pairs of variational input and output functions with compact support. A minimal control system is Hamiltonian if and only if for the system this symplectic form equals zero.

The explicit consideration of _variational_ input-output behaviors makes our approach rather distinct from other approaches to the Hamiltonian realization problem as have been taken by Crouch & Irving [C1, C2] and Jakubczyk [J1, J5]. Here one gives conditions in terms of the input-output map of the control system. For a discussion of these results, especially of the explicit conditions found by Jakubczyk, we refer to chapter (1).

Finally let us put the theory discussed above in a historical perspective. During the last century the main emphasis in the study of Hamiltonian and Lagrangian systems has completely shifted from the original equations of Newton, Hamilton and Lagrange, including external forces, to the equations with _zero_ external forces [Sa1, Sa2]. Consequently the scope of the theory, and especially of new developments such as the geometrization of Hamiltonian systems, has been narrowed to _analytical_ mechanics. Since in most mechanical systems external forces do play a prominent role this entails quite a loss of generality. We are of the opinion that the system theoretic approach to Hamiltonian and Lagrangian systems, as taken in this monograph, can again enlarge the scope of current theory by reintroducing external forces. Furthermore the Hamiltonian Realization Problem in particular seems to be capable of adding new insight into classical and quantum mechanical physics; we mention the theory of [V1] and the introduction of a generalized moment map in [J7] (for system theoretic applications of some familiar tools in geometric quantization theory see, for example, Crouch and Irving [C2], Goncalves [G1], and Bloch [B1]). Indeed we hope that this monograph will stimulate the cross-fertilization between physics and system theory.

## 1. THE HAMILTONIAN REALIZATION PROBLEM

The present chapter outlines the results needed to understand the significance and position of the results we present here in relation to known results on the Hamiltonian realization problem and the inverse problem in classical mechanics. This is especially necessary since at first sight the work by Jakubczyk [J1], [J2], [J5], [J7] and to a far less extent that by Crouch and Irving [C1], [C2], would seem to have completely solved the Hamiltonian realization problem, without any mention of the inverse problem. After describing these results we outline our main results in which we essentially validate the conjecture of Van der Schaft [V1], [V5], then review the remaining chapters of the monograph. Complete definitions and results are given as they are needed in the main body of the work.

## SECTION 1.1

The nonlinear systems we deal with fall roughly into two classes, the affine (in control) systems

$$\dot{x} = g_0(x) + \sum_{i=1}^{m} u_i g_i(x) \quad , \quad x(0) = x_0 \quad , \quad x \in M$$

(1.1)

$$y_j = H_j(x) \quad , \quad 1 \leq j \leq p, \quad u = (u_1, \cdots, u_m) \in \Omega \subset R^m$$

and the general nonlinear systems

$$\dot{x} = f(x,u) \quad , \quad x(0) = x_0, \quad x \in M$$

(1.2)

$$y_j = h_j(x,u) \quad , \quad 1 \leq j \leq p, \quad u = (u_1, \cdots, u_m) \in \Omega \subset R^m$$

Here M is a differentiable manifold, $g_i$, $0 \leq i \leq m$ are vector fields on M, for each $u \in \Omega$, $f(\cdot,u)$ is a vector field on M, $H_j$, $1 \leq j \leq p$ are real valued functions on M, and for each $u \in \Omega$, $h_j(\cdot,u)$ are real valued functions on M, $1 \leq j \leq p$. We will be mainly concerned with situations where all the above data is real analytic $(C^\omega)$ but for the present we also admit the situation where all data is smooth $(C^\infty)$. These systems will then be called analytic or smooth respectively. System (1.2) will be called jointly analytic, if f and $h_j$ are jointly analytic in x and u. Let U be the space of all piecewise constant right continuous $\Omega$ valued functions on $[0,\infty)$. An element $u \in U$ is called an input or control. Given such an element u, we may solve equations (1.1) or (1.2) on some interval $[0,\varepsilon)$. We shall suppose that the vector fields $g_i$, $0 \leq i \leq m$ and $f(\cdot,u)$, $u \in \Omega$, are complete, so that we may take $\varepsilon = \infty$. In this case we call the system complete. Thus if Y is the space of $R^p$ valued continuous functions on $[0,\infty)$, to each $u \in U$ we obtain a unique $y \in Y$, where $y(t) = (y_1(t), \cdots, y_p(t))$ is called the output or observation function. Each initialized system $\Sigma$ either (1.1) or (1.2) therefore yields a causal input-output map

$$\Phi_\Sigma(x_0) : U \longrightarrow Y$$

defined by $\Phi_\Sigma^j(x_0)(u)(t) = y_j(t)$, (causal refers to the fact that if $u$, $v \in U$ and $u(t) = v(t)$ for $t \in [0,T]$ then $\Phi_\Sigma(x_0)(u)(t) = \Phi_\Sigma(x_0)(v)(t)$ for $t \in [0,T]$). We shall simply write $\Phi$ when it is clear that we are refering to that input-output map arising from a system $\Sigma$ initialized at $x_0$; or when $\Phi$ is given without reference to a particular system.

The system $\Sigma$, either (1.1) or (1.2), may be viewed as an (internal) state space realization of the (external) input-output map $\Phi_\Sigma(x_0)$. The basic realization problem is now simply stated as: - given a causal mapping $\Phi : U \longrightarrow Y$ when does there exist an initialized system $\Sigma$ such that $\Phi_\Sigma(x_0) = \Phi$, with the restriction that $\Sigma$ must lie within some predetermined class of systems. Clearly we may choose many classes of system and each would have a corresponding realization theory. We concern ourselves here only with those classes of system which are described by equations (1.1) or (1.2), or suitable generalizations described in chapter (6). The global realization problem as described here has a pleasing resolution, as given in Jakubczyk [J1], [J2] for the class of analytic systems given by equations (1.2). We therefore use his terminology in the sequel. Let

$$(1.3) \qquad a = (\omega_1, t_1)(\omega_2, t_2) \cdots (\omega_k, t_k)$$

represent the piecewise constant function on $[0, T_k)$ defined by setting, $T_r = \sum_{i=1}^{r} t_i$, $t_i > 0$, and $a(t) = \omega_r \in \Omega$ for $t \in [T_{r-1}, T_r)$, $T_0 = 0$. Let $a\omega$ represent the function on $[0, T_k]$ which coincides with $a$ on $[0, T_k)$, and has value $\omega \in \Omega$ at $T = t_k$. The input-output map $\Phi = \Phi_\Sigma(x_0)$ is now completely determined by the totality of possible values assumed by

$$(1.4) \qquad \langle \Phi, a\omega \rangle = \Phi(a\omega)(T_k)$$

We note that dealing with systems (1.1), or (1.2) where $h_j(\cdot, u) = h_j(\cdot)$, $1 \le j \le p$, it suffices to consider the inputs $a$ in (1.3) alone, rather than the inputs $au$.

An input-output map $\Phi$ is said to be (jointly) analytic if all the following maps are analytic and have analytic continuations to all of $(R^k \times \Omega^{k+1})$, $R^k$,

$$((t_1, \cdots, t_k, \omega_1, \cdots, \omega_k, \omega_{k+1}) \longrightarrow \langle \Phi, (\omega_1, t_1) \cdots (\omega_k, t_k) \omega_{k+1} \rangle)$$

$$(t_1 \cdots t_k) \longrightarrow \langle \Phi, (\omega_1, t_1) \cdots (\omega_k, t_k) \omega_{k+1} \rangle \text{ respectively.}$$

$\Phi$ is said to have finite rank if rank $\Phi = \sup \, \text{rank} \, \{\frac{\partial}{\partial t_i} \langle \Phi^{q_j}, a_k \cdot b_j \omega_j \rangle\}_{i,j=1}^{k} < \infty$ where $a_k, b_1 \cdots b_k$ are piecewise constant inputs of the form given in (1.3), $a_k = (\omega_1, t_1) \cdots (\omega_k, t_k)$, $\omega_j \in \Omega$, and the supremum is taken over all possible $k \geq 1$, $a_k, b_j \omega_j$, and integers $q_j$, $1 \leq q_j \leq p$, $1 \leq j \leq k$. Note $a_k \cdot b_j \omega_j$ is just the concatenation of the two inputs.

THEOREM 1.1 JAKUBCZYK [J1]

A causal jointly analytic input-output map $\Phi$, with $\Omega$ compact and convex, has a jointly analytic and complete realization by a system (1.2) if and only if rank $\Phi$ is finite.                                                                                  □

The original paper by Jakubczyk [J2] deals with systems of the form (1.2) with $h_j(\cdot, u) = h_j(\cdot)$, $1 \leq j \leq p$. Slightly modified definitions are used, and inparticular analyticity with respect to $\omega \in \Omega$ and the conditions on $\Omega$ are removed, to obtain analytic systems rather than jointly analytic. There are significant difficulties in extending these results to the smooth case, unless significant restrictions are placed on the system, see Sussmann [S1], Hermann and Krener [H], and Gauthier and Bornard [Ga]; or new definitions are introduced, see Goncalves [G2]. Since these difficulties are basically of a global nature this does not rule out the existence of, even $C^k$, local realizations. Fliess [F1], [F2], works out the local theory for systems (1.1), with analytic data, and Jakubczyk [J3] works out the general $C^k$ local theory. Since our theory is largely of a global nature we do not give any details here.

Although theorem (1.1) and similar results demonstrate the existence of realizations, an important question, especially in our monograph, is that of the uniqueness of the realizations. In general there is no reason to believe any realization of a particular input-output map should be unique. However if we restrict to those realizations, in which the dimension of the state manifold M is minimized, there is a type of uniqueness. These realizations are called minimal realizations, and the existence and uniqueness of these minimal realizations is one of the primary goals of the papers [S1], [H], [Ga], [G2], although it is implicit in the work [J2], and all the local theory. In fact in the global situation the minimal realizations described above correspond more closely with the quasi-minimal realizations defined in the above references.

SECTION 1.2

The realizability conditions described in theorem (1.1) may also be interpreted in terms of series expressions for the input-output map, and these conditions are nicely reviewed in [J6]. The two main series representations of the input-output map are the power series in noncommuting variables, and the Volterra series. The power series approach initially relied on the fact that, at least for analytic systems (1.1), the input-output map may be defined by a convergent series

$$(1.5) \qquad y_{i_0}(t) = \alpha_{i_0}(x_0) + \sum_{\substack{i_1,\cdots,i_k \\ k \geq 1}} \alpha_{i_0 i_1 \cdots i_k}(x_0) \, \xi_{i_1 \cdots i_k}(t)$$

where $1 \leq i_0 \leq p$ , $0 \leq i_j \leq m$ , for $1 \leq j \leq k$ , and setting $u_0(t) \equiv 1$

$$\xi_i(t) = \int_0^t u_i(s)ds, \quad \xi_{i_1,\cdots,i_k}(t) = \int_0^t u_{i_k}(s) \, \xi_{i_1,\cdots,i_{k-1}}(s)ds,$$

$$\alpha_{i_0} = H_{i_0}, \quad \alpha_{i_0 i_1 \cdots i_k} = g_{i_k}(\alpha_{i_0 i_1 \cdots i_{k-1}}).$$

The latter expression is just the Lie derivative of $\alpha_{i_0 i_1 \cdots i_{k-1}}$ by the vector field $g_{i_k}$ . If $X_0, X_1, \cdots, X_m$ are noncommuting variables the corresponding power series is simply

$$\alpha_{i_0}(x_0) + \sum_{\substack{i_1,\cdots,i_k \\ k \geq 1}} \alpha_{i_0 i_1 \cdots i_k}(x_0) \, X_{i_0} X_{i_1} \cdots X_{i_k}$$

See Fliess [F1], [F2], for the (local) realization theory based on this representation.

The Volterra series representation of the input-output map of system (1.1) is basically a recombination of certain terms in (1.5). For $m = p = 1$, write $u_1(t) = u(t)$, $g_1(t) = g(t)$, $H_1(t) = H(t)$, to obtain

$$
\begin{aligned}
(1.6) \qquad y(t) &= W_0(t,x_0) + \int_0^t W_1(t,\sigma_1,x_0) \, u(\sigma_1)d\sigma_1 \\
&+ \int_0^t \int_0^{\sigma_1} W_2(t,\sigma_1,\sigma_2,x_0) \, u(\sigma_1) \, u(\sigma_2)d\sigma_1 d\sigma_2 + \cdots
\end{aligned}
$$

For analytic systems the series converges to the input-output map of (1.1), (in a suitable uniform sense) to the output function y. The following formula for the Volterra kernels $W_k(t,\sigma_1,\cdots,\sigma_k,x_0)$ developed in Leslak and Krener [L], clearly shows the analytic dependence of the kernels on their arguments.

(1.7)     $W_k(t,\sigma_1,\cdots,\sigma_k,x_0)$

$$= \gamma(-\sigma_k)_* \, g(\gamma(\sigma_k)x_0)(\gamma(-\sigma_{k-1})_* \, g(\gamma(\sigma_{k-1})\cdot))(\cdots$$

$$\cdots(\gamma(-\sigma_1)_* \, g(\gamma(\sigma_1)\cdot))(H\circ\gamma(t)\cdot)\cdots)$$

where $(t,x) \longrightarrow \gamma(t)x$ is the flow of the vector field $g_0$.
For systems (1.2) the following more general expansion is appropriate, see
Jakubczyk [J4]. We let $\Omega^*$ denote the free monoid generated by $\Omega$. Thus elements of
$\Omega$ are thought of as noncommuting variables, and $\Omega^*$ is the set of words
$\omega_1\omega_2\cdots\cdots\omega_k$ , $\omega_j \in \Omega$. A formal power series in $\Omega$ is just a map $P : \Omega^* \longrightarrow R$, but
we are usually interested in p series together, so we view $P$ as a map
$P : \Omega^* \longrightarrow R^p$. We write

$$P = \sum_{\omega\in\Omega^*} \langle P,\omega\rangle \, \omega$$

Given an analytic input-output map $\Phi : U \longrightarrow Y$ we define

(1.8)     $$\Phi_{\omega_1\cdots\cdots\omega_k} = \frac{\partial}{\partial t_1} \cdots\cdots \frac{\partial}{\partial t_{k-1}} \bigg|_{t_i = 0} \langle\Phi,(\omega_1,t_1)\cdots\cdots(\omega_{k-1},t_{k-1})\omega_k\rangle$$

Now $\Phi$ defines a formal power series

(1.9)     $$P = \sum_{\substack{\omega_1,\cdots,\omega_k\in\Omega \\ k\geq 1}} \Phi_{\omega_1\cdots\cdots\omega_k} \, \omega_1\cdots\cdots\omega_k$$

Conversely a formal power series

$$P = \sum_{\substack{\omega_1,\cdots\omega_k\in\Omega \\ k\geq 1}} \langle P,\omega_1\cdots\cdots\omega_k\rangle \, \omega_1\cdots\cdots\omega_k$$

such that each series

$$\sum_{i_1,\cdots,i_k} \langle P,\omega_1^{i_1}\cdots\cdots\omega_{k-1}^{i_{k-1}}\omega_k\rangle \frac{t_1^{i_1}\cdots\cdots t_{k-1}^{i_{k-1}}}{i_1!\cdots\cdots i_k!} = \Phi_{\omega_1\cdots\cdots\omega_k}(t_1\cdots\cdots t_{k-1})$$

converges in a neighbourhood of $0 \in R^k$, and has an analytic continuation to all of
$R^k$, defines an analytic input-output map $\Phi$, (here we use the notation

$$\omega_1^{i_1}\cdots\cdots\omega_{k-1}^{i_{k-1}} \, \omega_k = \overbrace{\omega_1\cdots\cdots\omega_1}^{i_1 \text{ times}}\cdots\cdots\overbrace{\omega_{k-1}\cdots\cdots\omega_{k-1}}^{i_{k-1} \text{ times}} \omega_k).$$ Indeed $\Phi$ is specified, as in
(1.4), by the values

$$\langle \Phi, (\omega_1, t_1) \cdots (\omega_{k-1}, t_{k-1}) \omega_k \rangle = \Phi_{\omega_1 \cdots \omega_k} (t_1 \cdots t_{k-1}).$$

If $\Phi = \Phi_\Sigma(x_0)$, $f^\omega(x) = f(x, \omega)$, $h_j^\omega(x) = h_j(x, \omega)$ then

$$(1.10) \qquad \Phi_{\omega_1 \cdots \omega_k}^j = f^{\omega_1} (f^{\omega_2} (\cdots (f^{\omega_{k-1}} (h_j^{\omega_k}) \cdots))(x_0)$$

Note the similarity between the coefficients (1.10) of the power series (1.9), and the coefficients of the expansion (1.5) in the affine case. In [J4] Jakubczyk works out a local realization theory, in the analytic case, for systems (1.2) in terms of the series (1.9) and operations on it.

SECTION 1.3

As already explained in the Introduction, our specific interest in this paper lies in systems (1.1) and (1.2) with added structure motivated by the equations of Lagrange and Hamilton, namely

$$(1.11) \qquad \frac{d}{dt} \left( \frac{\partial}{\partial \dot{q}_i} L(q, \dot{q}) \right) - \frac{\partial}{\partial q_i} L(q, \dot{q}) = F_i \quad , \quad 1 \leq i \leq n$$

and

$$(1.12) \qquad \dot{q}_i = \frac{\partial H}{\partial p_i}(q, p) \quad , \quad \dot{p}_i = -\frac{\partial H}{\partial q_i}(q, p) + F_i \quad , \quad 1 \leq i \leq n.$$

Here we include external forces $F_i$, as in the original conception by Lagrange and Hamilton, see Santilli [Sa2], and $q = (q_1, \cdots, q_n)$, $\dot{q} = (\dot{q}_1, \cdots, \dot{q}_n)$ or $q = (q_1, \cdots, q_n)$, $p = (p_1, \cdots, p_n)$ lie in some open subset $U$ of $R^{2n}$. We notice that if the matrix with components $\left[ \dfrac{\partial^2 L}{\partial \dot{q}_i \partial \dot{q}_j} \right]_{i,j=1}^n$ is nonsingular on $U$, we may use the Legendre transform to rewrite equations (1.11) in the form of equations (1.12). Since we prefer to work with first order differential equations we shall therefore concentrate on equations (1.12).

In the systems theoretic setting we view the external forces $F_i$ as controls, or inputs, and write $u_i = F_i$, $1 \leq i \leq n$. If we also assume that only the first $m$ components $F_1, \cdots, F_m$ are non zero, equations (1.12) become

$$\dot{q}_i = \frac{\partial H_0}{\partial p_i}(q, p) \qquad , \quad 1 \leq i \leq n$$

$$(1.13) \qquad \dot{p}_i = -\frac{\partial H_0}{\partial q_i}(q, p) + u_i \quad , \quad 1 \leq i \leq m$$

$$\dot{p}_i = -\frac{\partial H_0}{\partial q_i}(q,p) \qquad , \quad m+1 \le i \le n$$

This situation may be generalized (see the Introduction) by introducing m observation, or output maps $H_j : U \longrightarrow R$ , $1 \le j \le m$ and defining the system

$$(1.14) \qquad \begin{bmatrix} \dot{q}_i \\ \dot{p}_i \end{bmatrix} = \begin{bmatrix} \dfrac{\partial H_0}{\partial p_i} \\ -\dfrac{\partial H_0}{\partial q_i} \end{bmatrix} - \sum_{j=1}^{m} u_j \begin{bmatrix} \dfrac{\partial H_j}{\partial p_i} \\ -\dfrac{\partial H_j}{\partial q_i} \end{bmatrix} , \quad 1 \le i \le n$$

$$y_j = H_j(q,p) \quad , \quad 1 \le j \le m$$

Notice that by setting $H_j(q,p) = q_j$ we recover the system (1.13). The idea in this generalization is that we can exert external forces in the directions determined by the observations we make. See the Introduction and Van der Schaft [V1], [V2], [V3], [V5], Willems and Van der Schaft [W2] and Brockett [B1], for further discussion concerning the introduction of systems such as (1.14).

The equations (1.14) lead to the following coordinate free definition of an affine (locally) Hamiltonian system ([V1], [V3], [V4])

$$(1.15) \qquad \dot{x} = g_0(x) - \sum_{i=1}^{m} u_i X_{H_i}(x) \quad , \quad x(0) = x_0.$$

$$y_i = H_i(x) \ , \ 1 \le i \le m \ , \ x \in (M,\omega) \ , \ u \in \Omega \subset R^m$$

where $(M,\omega)$ is a symplectic manifold ([A]), with symplectic form $\omega$, $X_{H_i}$ are globally Hamiltonian vector fields determined from the relationships

$$\omega(X_{H_i},\cdot) = -dH_i \ , \ 1 \le i \le m$$

and $g_0$ is an infinitesimally symplectic vector field. The latter condition just states that $L_{g_0}\omega = 0$ where $L_{g_0}$ is the Lie derivative, so it follows ([A]) that $g_0$ is a locally Hamiltonian vector field; that is locally there is a function $H_0$ such that

$$\omega(g_0,\cdot) = -dH_0.$$

If $g_0$ is a globally Hamiltonian vector field we say that system (1.15) is an affine globally Hamiltonian system. In any case the Darboux theorem ([A]) shows that about any point in $(M,\omega)$ there are local coordinates $(q_1,\cdots,q_n,p_1,\cdots,p_n)$ for M in which $\omega = \sum_{i=1}^{n} dp_i \wedge dq_i$. It follows that in these coordinates, system

(1.15) is indeed given by the equations (1.14).

Although most of our work concerns Hamiltonian systems in the form of equations (1.15) we also wish to consider more general Hamiltonian systems, corresponding to general nonlinear systems (1.2). Thus a general (globally) Hamiltonian system is defined by the equations

(1.16)
$$\dot{x} = X_H(x,u) \qquad , \quad x(0) = x_0 \quad , \quad x \in (M,\omega)$$

$$y_i = c_i \frac{\partial H}{\partial u_i}(x,u) \quad , \quad 1 \le i \le m, \quad u \in \Omega \subset R^m, \quad c_i = \pm 1,$$

where $(M,\omega)$ is a symplectic manifold, and for each $u \in \Omega$, $X_H(x,u)$ is a Hamiltonian vector field with Hamiltonian function $H(\cdot,u)$. However this definition is not quite general enough. Work in [W], [V1], [V2], [V5] and [B1] demonstrate this point, and inparticular show that in many physical systems it is not possible to make a global distinction between input variables $u_i$ and output variables $y_i$. Instead all the external variables are combined, and assumed to evolve in some other manifold W. This gives rise to the general definition of a nonlinear system which, in local coordinates $(y,u)$ for W, takes the form of equations (1.2). Similarly we define a general (locally) Hamiltonian system in which $(M,\omega)$ and $(W,\omega^e)$ are symplectic manifolds, such that in local (Darboux) coordinates for W and M the system coincides with the local coordinate version of system (1.16), namely

(1.17)
$$\dot{q}_i = \frac{\partial H}{\partial p_i} (q,p,u) \qquad , \qquad 1 \le i \le n$$

$$\dot{p}_i = - \frac{\partial H}{\partial q_i} (q,p,u) \qquad , \qquad 1 \le i \le n$$

$$y_i = c_i \frac{\partial H}{\partial u_i} (q,p,u) \qquad , \qquad 1 \le i \le m , \qquad c_i = \pm 1.$$

The Hamiltonian realization problem is a special case of the general problem defined in section (1.1), where we seek realizations within a class of Hamiltonian systems described above. The importance of minimal realizations of input-output maps is now evident. Without the minimality constraint, even if an input-output map has a Hamiltonian realization, there will exist many non Hamiltonian realizations, in the general class of nonlinear systems described above. However as is already known, see Goncalves [G1], and Van der Schaft [V4], if an input-output map has a (globally) Hamiltonian quasi-minimal realization, then any other quasi-minimal realization differing by a (global) change of coördinates, is in fact also (globally) Hamiltonian, and the two system differ by a symplectic change of coordinates. Thus to a large extent uniqueness of the Hamiltonian structure is guaranteed. In fact this situation is completely resolved in our work here, but see also the work by Jakubczyk [J1], [J5], [J7].

Remark (see [J1]) Notice that without the restriction to natural outputs as in (1.15) or (1.17) the Hamiltonian realization problem is much less well-defined. In fact, any nonlinear input-output map with a realization $\dot{x} = f(x,u)$, $y = h(x,u)$, $x \in M$, also admits the pseudo-Hamiltonian realization

$$\dot{x} = \frac{\partial H}{\partial p} (x,p,u)$$

$$y = h(x,u)$$

$$\dot{p} = - \frac{\partial H}{\partial p} (x,p,u)$$

with $H(x,p,u) = p^T f(x,u)$. Of course such a realization is not minimal.

## SECTION 1.4

Although the previously cited work [G1] and [V1], [V4], deals with existence and uniqueness of (quasi) minimal Hamiltonian realizations, it does not contain any realizability conditions. The original result in this area is as follows.

THEOREM 1.2 BROCKETT AND RAHIMI [B2]

The input-output map

$$y(t) = \int_0^t W(t,\sigma)u(\sigma)d\sigma$$

of the linear system $\dot{x} = Ax + Bu$, $y = Cx$, $x(0) = 0$, $m = p$, has a linear Hamiltonian realization if and only if

(1.18)     $W(t,\sigma) = -W(\sigma,t)^T$                                    □

This result was generalized to systems with finite Volterra series as follows. Define, as in Crouch [C4], the following bracket operation on the Volterra kernels, for the case $m = p = 1$,

$$W_k(t,\sigma_1,\cdots,[\sigma_i,\sigma_{i+1}],\cdots,\sigma_k) = W_k(t,\sigma_1,\cdots,\sigma_i,\sigma_{i+1},\cdots,\sigma_k)$$
$$- W_k(t,\sigma_1,\cdots,\sigma_{i+1},\sigma_i,\cdots,\sigma_k).$$

and hence inductively via

$$W_k(t,\sigma_1,\cdots,[\sigma_i,\cdots,\sigma_j],\cdots,\sigma_k) = W_k(t,\sigma_1,\cdots,[\sigma_i,\cdots,\sigma_{j-1}],\sigma_j,\cdots,\sigma_k)$$

(1.19)

$$- W_k(t,\sigma_1,\cdots,\sigma_{i-1},\sigma_j,[\sigma_i,\cdots,\sigma_{j-1}],\sigma_{j+1},\cdots,\sigma_k).$$

THEOREM 1.3 CROUCH AND IRVING [C1].

An input-output map $\Phi$ which has a realization by an analytic, affine and complete system (1.1) in which $g_0(x_0) = 0$, and has a representation as a finite Volterra series of length N, has a realization by an affine, analytic and complete globally Hamiltonian system (1.15) if and only if

(1.20)  $$W_k([t,\sigma_1,\cdots,\sigma_r],\cdots,\sigma_k) = (r + 1) W_k(t,\sigma_1,\cdots,\sigma_k)$$

for $1 \le r \le k$ and $1 \le k \le N$.  □

A similar result is obtained if $g_0(x_0) \ne 0$, but then time varying Hamiltonian system are required. See Goncalves [G1] for the best available exposition of time varying Hamiltonian realization theory. We defer any discussion of this result until we review some recent work of Jakubczyk, [J1], [J5], [J7]. Indeed he considers both the global and local theory, but we only consider the global case here. Jakubczyk first introduces another class of systems related to these defined in equations (1.16), namely those defined by equations of the form

(1.21)
$$\dot{x} = X_H(x,u) \quad , \quad x(0) = x_0 \ , \ x \in (M,\omega)$$

$$y = H(x,u) \quad , \quad u \in \Omega \subset R^m.$$

where again $(M,\omega)$ is a symplectic manifold and $X_H(\cdot,u)$ is the Hamiltonian vector field with Hamiltonian $H(\cdot,u)$ for each $u \in \Omega$. Note the output space for these system is always R. To state the first result we must also introduce more notation related to the quantities $\Phi_{\omega_1\cdots\cdots\omega_k}$ defined in equation (1.8) for an input-output map $\Phi$, and similar to that in (1.19). Let

$$\Phi_{\omega_1\cdots\cdots[\omega_i,\omega_{i+1}]\cdots\cdots\omega_k} = \Phi_{\omega_1\cdots\cdots\omega_i\omega_{i+1}\cdots\cdots\omega_k}$$

$$- \Phi_{\omega_1\cdots\cdots\omega_{i+1}\omega_i\cdots\cdots\omega_k}$$

and define inductively

(1.22)
$$\Phi_{\omega_1 \cdots \cdot [\omega_i \cdots \cdot \omega_j] \cdots \cdot \omega_k} = \Phi_{\omega_1 \cdots \cdot \omega_i [\omega_{i+1} \cdots \omega_j] \cdots \cdot \omega_k}$$
$$- \Phi_{\omega_1 \cdots \cdot \omega_{i-1} [\omega_{i+1} \cdots \cdot \omega_j] \omega_i \cdots \cdot \omega_k}.$$

THEOREM 1.4 JAKUBCZYK [J1], [J5]

A causal, jointly analytic input-output map $\Phi$, with $\Omega$ compact and convex, has a jointly analytic complete realization by a system (1.21) if and only if the rank of $\Phi$ is finite and

(1.23)
$$\Phi_{[\omega_1 \cdots \cdot \omega_k]} = k \, \Phi_{\omega_1 \cdots \cdot \omega_k}, \quad k \geq 2, \ \omega_i \in \Omega, \ 1 \leq i \leq k$$

□

To overview the proof of this result note that without the conditions (1.23) we obtain a minimal realization in the form of a system (1.2) by theorem (1.1). Since in this case $p = 1$ set $h_j = h$, and $h_{\omega_1 \cdots \cdot \omega_k} = f^{\omega_1}(f^{\omega_2} \cdots \cdot (f^{\omega_{k-1}}(h^{\omega_k}) \cdots \cdot) \cdots \cdot)$ as in equation (1.10). Now (1.23) implies that

(1.24)
$$h_{[\omega_1 \cdots \cdot \omega_k]}(x_0) = k \, h_{\omega_1 \cdots \cdot \omega_k}(x_0)$$

for $k \geq 2$, $\omega_i \in \Omega$, $1 \leq i \leq k$, where the definition of $h_{[\omega_1 \cdots \cdot \omega_k]}$ is analogous to that of $\Phi_{[\omega_1 \cdots \cdot \omega_k]}$. The following intermediate result is of independent interest and should be compared with our theorem 4.2 of chapter 4.

THEOREM 1.5 JAKUBCZYK [J1], [J5].

A minimal analytic system (1.2) satisfies the identities (1.24) if and only if the identities are satisfied identically on M, and if and only if M is a symplectic manifold, $f(\cdot, u)$ is a Hamiltonian vector field $X_H(\cdot, u)$ with Hamiltonian $H(\cdot, u)$ for each $u \in \Omega$.

□

We see immediately that our system (1.2) can be rewritten in the form of system (1.21) as desired. To motivate the conditions (1.23) we recall the definition of Poisson bracket on a symplectic manifold $(M, \omega)$, between two functions $\alpha$ and $\beta$,

$$\{\beta, \alpha\} = \omega(X_\beta, X_\alpha) = X_\beta(\alpha)$$

where $X_\alpha$ and $X_\beta$ are the associated Hamiltonian vector fields. Thus in the situation of theorem (1.5) we have

$$f^{\omega_1}(h^{\omega_2}) = X_H(\cdot,\omega_1) \ H(\cdot,\omega_2) = \{H(\cdot,\omega_1), \ H(\cdot,\omega_2)\} = \{h^{\omega_1}, h^{\omega_2}\}$$

and more generally

$$h_{\omega_1 \cdots \omega_k} = \{h^{\omega_1}, \{h^{\omega_2}, \{\cdots \cdot \{h^{\omega_{k-1}}, h^{\omega_k}\} \cdots \cdot \}.$$

The fact that $h_{\omega_1 \cdots \omega_k}$ must satisfy the conditions of theorem (1.5) is now just a direct result of another result by Dynkin-Specht-Wever, which characterizes Lie monomials in noncommuting variables. See Ree [R] for a nice exposition of this result, and Lamnabhi-Lagarrigue and Crouch [La] for further implications for series representations of input-output maps. This result also explains the analogous conditions (1.20) on the Volterra kernels. Indeed the expressions (1.7) for the Volterra kernels may be rewritten for a Hamiltonian system (1.15), in the form

$$W_k(t,\sigma_1, \cdots, \sigma_k, x_0) = \{g(\sigma_k), \{g(\sigma_{k-1}), \cdots \cdot \{g(\sigma_1), g(t)\} \cdots \cdot \}(x_0)$$

where $g(\sigma) = H \circ Y(\sigma)$. (We use the terminology introduced for the expression (1.7).) To obtain a realizability result for systems of the form (1.16) rather than (1.21) Jakubczyk proceeds as follows. Given an input-output map $\phi$ define a causal mapping $\tilde{\phi} : U \longrightarrow \tilde{Y}$, where $\tilde{Y}$ is now just the continuous functions on $[0,\infty)$, by setting

(1.25) $\qquad \langle \tilde{\phi}, a_k \omega_{k+1} \rangle = \sum\limits_{s=1}^{k} \int\limits_{\omega_s}^{\omega_{s+1}} \sum\limits_{i=1}^{m} \langle \phi^i, a_s u \rangle \ du_i$

where $a_s = (\omega_1,t_1) \cdots \cdot (\omega_s,t_s)$. That this is well defined, i.e. that the integrals are independent of path, requires that the components $\phi^i$ satisfy

(1.26) $\qquad \dfrac{\partial}{\partial u_i} \langle \phi^j, a_k u \rangle = \dfrac{\partial}{\partial u_j} \langle \phi^i, a_k u \rangle, \ 1 \le i, j \le m.$

Another intermediate result states that an analytic input-output map $\phi$ has a realization by a Hamiltonian system (1.16) if and only if $\tilde{\phi}$ has a realization by a system of the form given in (1.21). Necessity is clear since

$$\dfrac{\partial}{\partial u} \langle \tilde{\phi}, a_k u \rangle = \langle \phi, a_k u \rangle.$$

Conversely if $\phi$ has a realization by a Hamiltonian system (1.16) then $\langle \phi^i, a_j u \rangle = \dfrac{\partial H}{\partial u_i}(x(T_j),u)$ where $T_j = \sum\limits_{i=1}^{j} t_i$, $a_j = (\omega_1,t_1) \cdots \cdot (\omega_j,t_j)$ as before. Inparticular (1.26) is satisfied and we obtain from (1.25)

$$\langle \tilde{\phi}, a_k \omega_{k+1} \rangle = \sum_{s=0}^{k} \left( H(x(T_s), \omega_{s+1}) - H(x(T_s), \omega_s) \right).$$

But $H(x(T_s), \omega_s) = H(x(T_{s-1}), \omega_s)$ since $H(\cdot, \omega_s)$ is constant along the trajectory $x(t)$, $t \in [T_{s-1}, T_s)$, which is governed by the Hamiltonian vector field $X_H(\cdot, \omega_s)$ by construction. Thus

$$\langle \tilde{\phi}, a_k \omega_{k+1} \rangle = H(x(T), \omega_{k+1}) - H(x_0, u_0)$$

Since $H(x_0, u_0)$ is just a constant we see that $\tilde{\phi}$ is the input-output map of system (1.21). Combining this result with theorem (1.4) the following result is obtained.

THEOREM 1.6   JAKUBCZYK [J1], [J5]

A causal jointly analytic input-output map $\phi$, with $\Omega$ compact and convex, has a jointly analytic complete realization by a system (1.16), if and only if the rank of $\phi$ is finite, $\phi$ satisfies (1.26), and $\tilde{\phi}_{\omega_1, \cdots, \omega_k}$ satisfies (1.23) where for $\tilde{\phi}$ given by (1.25)

$$\tilde{\phi}_{\omega_1, \cdots, \omega_k} = \frac{\partial}{\partial t_1} \cdots \cdot \frac{\partial}{\partial t_{k-1}} \Big|_{t_i = 0} \langle \tilde{\phi}, a_k \omega_k \rangle. \qquad \Box$$

The only work remaining in this result, after theorem (1.4) has been established, is to show that $\tilde{\phi}$ is of finite rank if and only if $\phi$ is also. In Jakubczyk [J5], [J7], [J8], a realization theory is worked out, for systems (1.21), based on the formal power series representation of the input-output map introduced in section (1.2). This theory which introduces a generalized moment map, is an "infinite dimensional" version of an idea in Goncalves [G1]. Note that the coefficients $\phi_{\omega_1, \cdots, \omega_k}$ in the power series (1.9) are exactly those that appear in the conditions (1.23) of theorem (1.4).

SECTION 1.5

The inverse problem in classical mechanics in its original form concerns the system of equations without external forces

$$(1.27) \qquad R_i(q, \dot{q}, \ddot{q}) = 0 \quad , \quad 1 \le i \le n \quad , \quad q \in \mathbb{R}^n$$

where $q = (q_1, \cdots, q_n)$ , $\dot{q} = (\frac{dq_1}{dt}, \cdots, \frac{dq_n}{dt})$ and $\ddot{q} = (\frac{d^2 q_1}{dt^2}, \cdots, \frac{d^2 q_n}{dt^2})$. We additionaly assume that the matrix $\left(\dfrac{\partial R_i}{\partial q_j}\right)^n_{i,j=1}$ is nonsingular on an open domain $U \subset R^{3n}$, and $R_i$ is $C^1$ on U. One asks under what further conditions on $R_i$ there exists a function $L(q,\dot{q})$ such that after a possible reordering of the indices we have

$$(1.28) \qquad \frac{d}{dt}\left(\frac{\partial L}{\partial \dot{q}_i}(q,\dot{q})\right) - \frac{\partial L}{\partial q_i}(q,\dot{q}) = R_i(q,\dot{q},\ddot{q}) \quad , \quad 1 \le i \le n.$$

The solution of this problem as detailed in Santilli [Sa1], [Sa2], along with many related topics, involves the following definitions and constructions. Let $q : I \longrightarrow R^n$ be a $C^2$ solution of the equations (1.27), on some open interval $I \subset R$, and define $r : I \longrightarrow U \subset R^{3n}$ by $r(t) = (q(t),\dot{q}(t),\ddot{q}(t))$. A variation of $q$ is a map $(t,\varepsilon) \longrightarrow q(t,\varepsilon)$, $I \times V \longrightarrow R^n$ where V is an open neighbourhood of 0 in R, such that $q(t,0) = q(t)$ for $t \in I$, and if $r(t,\varepsilon)$ is defined by $r(t,\varepsilon) =$ $= \left(q(t,\varepsilon), \dfrac{\partial q}{\partial t}(t,\varepsilon), \dfrac{\partial^2 q}{\partial t^2}(t,\varepsilon)\right)$ then $t \longrightarrow \dfrac{\partial r}{\partial \varepsilon}(t,0)$ exists and is continuous. Write $\dot{q}(t,\varepsilon) = \dfrac{\partial q}{\partial t}(t,\varepsilon)$, $\ddot{q}(t,\varepsilon) = \dfrac{\partial^2 q}{\partial t^2}(t,\varepsilon)$, and $\delta q(t) = \dfrac{\partial}{\partial \varepsilon}q(t,0)$, $\delta\dot{q}(t) = \dfrac{\partial}{\partial \varepsilon}\dot{q}(t,0)$, $\delta\ddot{q}(t) = \dfrac{\partial}{\partial \varepsilon}\ddot{q}(t,0)$. $\delta q$ is called the variational field along q. For notational convenience we set $\delta r(t) = \left(\delta q(t), \delta\dot{q}(t), \delta\ddot{q}(t)\right)$.

If $q(t,\varepsilon)$ is a variation of a solution $q(t)$ of (1.27), such that for each $\varepsilon \in V$ $t \to q(t,\varepsilon)$ is a $C^2$ solution of (1.27) then

$$R_i\left(q(t,\varepsilon), \dot{q}(t,\varepsilon), \ddot{q}(t,\varepsilon)\right) = 0, \qquad 1 \le i \le n.$$

It follows that by differentiating with respect to $\varepsilon$ we obtain the so called <u>variational</u> equations

$$(1.29) \qquad M\left(r(t)\right) \delta r(t) = 0$$

where M is a $n \times 3n$ matrix depending on r and consisting of partial derivatives of the components $R_i$. As in Santilli [Sa1] there exists a unique $n \times 3n$ matrix $M^*$, also depending on r, with the property that given two variations $q_1(t,\varepsilon)$ and $q_2(t,\varepsilon)$ of a solution $q(t)$ of (1.27), with corresponding variational fields $\delta_1 q$, $\delta_2 q$, there exists a unique function $Q(r,\delta_1 r,\delta_2 r)$ satisfying

$$(1.30) \qquad \delta_2^T q(t)M(r(t))\delta_1 r(t) - \delta_1^T q(t)M^*(r(t))\delta_2 r(t) = \frac{d}{dt} Q\left(r(t),\delta_1 r(t),\delta_2 r(t)\right).$$

Note that Q is bilinear in $(\delta_1 r,\delta_2 r)$. The equations

$$(1.31) \qquad M^*\left(r(t)\right) \delta r(t) = 0$$

are called the <u>adjoint variational</u> equations, and are well treated in many text-books on differential equations (see e.g. [Mo]). The desired result is now described in the following

THEOREM 1.7 (see [Sa1])

The inverse problem in classical mechanics has a solution if and only if the variational equation is self adjoint i.e. $M(r) = M^*(r)$ for any $r$.

The resulting conditions in terms of the functions $R_i(q,\dot{q},\ddot{q})$, $i = 1,\cdots,n$, are known as the Helmholtz conditions [Sa1]. We would like to point out that one of the essential constructions in Santilli's proof of sufficiency is the construction of a symplectic form, from which a Hamiltonian and hence Lagrangian structure is given to the equations. In our theory, especially theorem (4.2), this construction is essentially repeated showing that Santilli's construction is even more general than perhaps was appreciated.

As noted in Takens [T] and in more detail in Van der Schaft [V1], [V5], this result also solves a restricted Hamiltonian realization problem. Consider first a system of Newtonian equations

$$R_i(q,\dot{q},\ddot{q}) = u_i \quad , \quad 1 \le i \le m$$

(1.32)

$$R_i(q,\dot{q},\ddot{q}) = 0 \quad , \quad m+1 \le i \le n$$

under the same restrictions as imposed on equations (1.27). We ask when it is possible to write this as a set of Lagrangian or Hamiltonian equations with external forces? Clearly the conditions are exactly the same as in theorem (1.6), since once we have equations (1.28) we also know that

$$\frac{d}{dt} \frac{\partial L}{\partial \dot{q}_i} - \frac{\partial L}{\partial q_i} = u_i \quad , \quad 1 \le i \le m$$

$$\frac{d}{dt} \frac{\partial L}{\partial \dot{q}_i} - \frac{\partial L}{\partial q_i} = 0 \quad , \quad m+1 \le i \le n$$

As noted in section (1.2), if we are to write these equations in the form of a Hamiltonian system (1.13) the natural outputs, or observations, are $y_i = q_i$ $1 \le i \le m$. Thus adding these equations to those of (1.31), we see that theorem (1.6) provides us with an exact solution to the corresponding Hamiltonian realization problem.

Note however that we have not solved the problem by examining the properties of the corresponding input-output map directly, as is the case in theorems (1.2),

(1.3) and (1.5). Rather the problem is solved in terms of properties of variations in the state trajectories. In the case of system (1.31) with outputs $y_i = q_i$, each variation of the input function u, yields a corresponding variation in the state and output trajectories x and y, via equations the form

$$R_i\left(q(t,\varepsilon),\ \dot{q}(t,\varepsilon),\ \ddot{q}(t,\varepsilon)\right) = u_i(t,\varepsilon), \qquad 1 \leq i \leq m$$

$$R_i\left(q(t,\varepsilon),\ \dot{q}(t,\varepsilon),\ \ddot{q}(t,\varepsilon)\right) = 0\ , \qquad m+1 \leq i \leq n$$

$$y_i(t,\varepsilon) = q_i(t,\varepsilon)\ , \qquad 1 \leq i \leq m$$

$$u(t,0) = u(t),\ y(t,0) = y(t),\ q(t,0) = q(t).$$

It is assumed that the variations $u(t,\varepsilon)$ are such that the corresponding variational fields

$$\delta r(t),\ \delta u(t) = \frac{\partial u}{\partial \varepsilon}(t,0),\ \delta y(t) = \frac{\partial}{\partial \varepsilon} y(t,0)$$

exist and are continuous. By differentiating with respect to $\varepsilon$ we obtain

$$\left[M\left(r(t)\right)\delta r(t)\right]_i = \delta u_i(t) \qquad ,\quad 1 \leq i \leq m$$

$$\left[M\left(r(t)\right)\delta r(t)\right]_i = 0 \qquad ,\quad m+1 \leq i \leq n$$

$$\delta y_i(t) = \delta q_i(t) \qquad ,\quad 1 \leq i \leq m$$

In case the variational system is self adjoint equation (1.30) can now be written as

(1.33) $\qquad \delta_2^T y(t)\delta_1 u(t) - \delta_1^T y(t)\delta_2 u(t) = \frac{d}{dt} Q\left(r(t),\delta_1 r(t),\delta_2 r(t)\right)$

A generalization of this result for general Hamiltonian systems has already been given in Van der Schaft [V1], [V2], [V5]. In this paper we generalize equations (1.30) themselves, for general nonlinear systems, see lemma (2.1).

Now if $\delta_i u$ and $\delta_i y$, $i = 1,2$, have compact support in $(-\infty,\infty)$, then it is clearly true, given sufficient differentiability, that $(\delta_i q_j,\delta_i \dot{q}_j,\delta_i \ddot{q}_j)$, $1 \leq j \leq m$, have compact support. There is sufficient reason to believe that under appropriate conditions, and inparticular $m = n$, $\delta_i r$, $i = 1,2$ also have compact support. In this case equation (1.33) yields

(1.34)   $\int_{-\infty}^{\infty} (\delta_2^T y(t) \delta_1 u(t) - \delta_1^T y(t) \delta_2 u(t)) dt = 0$

Before further investigating equation (1.34) in the next section, we would like to mention an open problem which will not be dealt with in this monograph. This problem originates from the following generalization of the inverse problem in classical mechanics. Notice that the set of solutions of

(1.35)   $R_i(q, \dot{q}, \ddot{q}) = 0$     $i = 1, \cdots, n$

is not changed by pre-multiplication of this set of equations by a non-singular matrix $(\beta_{ij}(q, \dot{q}))_{i,j=1}^n$.

Hence given the equations (1.35) the question can be asked: When does there exist a non-singular multiplier matrix $\beta_{ij}(q, \dot{q})$ and a Lagrangian $L(q, \dot{q})$ such that

(1.36)   $\sum_{j=1}^{n} \beta_{ij}(q, \dot{q}) R_j(q, \dot{q}, \ddot{q}) = \frac{d}{dt}(\frac{\partial L}{\partial \dot{q}_i}) - \frac{\partial L}{\partial q_i},$     $i = 1, \cdots, n?$

An excellent discussion of this problem is given in [Sar], from which it is clear that finding explicit conditions for this problem seems very hard in general.

In the framework of Hamiltonian realization theory the problem amounts to the following. Given a control system (1.1), when does there exist a non-singular transformation of the inputs

(1.37)   $u_j = \sum_{k=1}^{m} \beta_{jk}(x) v_k$     $j = 1, \cdots, m$

with $v_1, \cdots, v_m$ the transformed inputs, such that the transformed system

(1.38)
$$\dot{x} = g_0(x) + \sum_{k=1}^{m} v_k (\sum_{j=1}^{m} \beta_{jk}(x) g_j(x))$$

$$y_j = H_j(x)     j = 1, \cdots, m     x(0) = x_0$$

is Hamiltonian? From a system theoretic point of view this suggests the even more general question: when does there exist a feedback

(1.39)   $u_j = \alpha_j(x) + \sum_{k=1}^{m} \beta_{jk}(x) v_k$ , $(\beta_{jk}(x))_{j,k=1}^m$   non-singular

such that the feedback transformed system

(1.40)
$$\dot{x} = g_0(x) + \sum_{j=1}^{m} \alpha_j(x) g_j(x) + \sum_{k=1}^{m} v_k (\sum_{j=1}^{m} \beta_{jk}(x) g_j(x))$$

$$y_j = H_j(x)     j = 1, \cdots, m$$

is Hamiltonian? This open problem fits in very well into current research of
finding normal forms for nonlinear systems by applying feedback and coordinate
transformations; see in this context also [St].

## SECTION 1.6

Before we state the conjecture of Van der Schaft, motivated by the observations of
section 1.5, it is clear that we must consider the more general situation of non
initialized systems (i.e. x(0) is arbitrary).

Given a complete system $\Sigma$, described by equations (1.2) we say that the behaviour
of the system $\Sigma_i$ is the set of time responses

$$t \longrightarrow \bigl(u(t),y(t),x(t)\bigr) \quad , \quad R \longrightarrow \Omega \times R^p \times M$$

satisfying the equations

(1.41) $\qquad \frac{dx}{dt}(t) = f\bigl(x(t),u(t)\bigr) , \; y(t) = h\bigl(x(t),u(t)\bigr) , \; t \in R$

and such that u is right continuous and piecewise constant. Our completeness
assumption ensures that this definition makes sense. The external behaviour $\Sigma_e$ is
just projection of $\Sigma_i$ into the set of input and output responses $t \longrightarrow \bigl(u(t),y(t)\bigr)$.
Let $\Sigma_i^+(T)$, $\bigl(\Sigma_e^+(T)\bigr)$ be the time responses obtained from $\Sigma_i(\Sigma_e)$ by restriction to
$[T,\infty)$. Because of the time invariance of the defining equations (1.40), $\Sigma_i^+(T_1)$
$\bigl(\Sigma_e^+(T_1)\bigr)$ differs from $\Sigma_i^+(T_2)$ $\bigl(\Sigma_e^+(T_2)\bigr)$ only by time translation. $\Sigma_i^+(T)$ $\bigl(\Sigma_e^+(T)\bigr)$ is a
union of subsets $\Sigma_i^+(T)(x_T)$ $\bigl(\Sigma_e^+(T)(x_T)\bigr)$ corresponding to those responses satisfying
$x(T) = x_T$. $\bigl(\Sigma_e^+(T)(x_T)$ is just the projection of $\Sigma_i^+(T)(x_T)$.$\bigr)$ Note that
$\Sigma_e^+(0)(x_0)$ may be identified with the input-output map $\Phi_\Sigma(x_0)$.

We define a variation of an element $(\bar{u},\bar{y},\bar{x}) \in \Sigma_i$, in the same fashion as before,
as a mapping $(t,\varepsilon) \longrightarrow \bigl(u(t,\varepsilon),y(t,\varepsilon),x(t,\varepsilon)\bigr)$, $R \times V \longrightarrow \Omega \times R^p \times M$ satisfying,
for each $\varepsilon \in V$ $t \longrightarrow \bigl(u(t,\varepsilon),y(t,\varepsilon),x(t,\varepsilon)\bigr) \in \Sigma_i$, and $\bigl(u(t,0),y(t,0),x(t,0)\bigr) =$
$= \bigl(\bar{u}(t),\bar{y}(t),\bar{x}(t)\bigr)$. Moreover we assume the corresponding variational field
$t \longrightarrow \bigl(\delta u(t),\delta y(t),\delta x(t)\bigr)$ exists, and that $\delta u$ is piecewise constant, $\delta y$ is
continuous and $\delta x$ is absolutely continuous. We define variations of elements in
$\Sigma_e$ by projection, and those in $\Sigma_i^+(T)$, $\Sigma_e^+(T)$ similarly. Variations of elements in
$\Sigma_i^+(T)(x_T)$ $\bigl(\Sigma_e^+(T)(x_T)\bigr)$ must satisfy an extra constraint $x(T,\varepsilon) = x_T$ for $\varepsilon \in V$, so
in this case $\bigl(\delta u(T),\delta y(T),\delta x(T)\bigr) = 0$.

A weakened version of the conjecture by Van der Schaft is as follows.

CONJECTURE 1.8 VAN DER SCHAFT [V1], [V5].

If $\Sigma_e$ represents the external behaviour of a general nonlinear system, then the system is Hamiltonian if and only if given any element $(\bar{u},\bar{y}) \in \Sigma_e$, any two variations $(\delta_i u, \delta_i y)$, $i = 1, 2$, of $(\bar{u},\bar{y})$, such that $(\delta_i u, \delta_i y)$ have compact support, satisfy

$$\int_{-\infty}^{\infty} (\delta_2^T y(t)\delta_1 u(t) - \delta_1^T y(t)\delta_2 u(t))dt = 0$$

☐

Although this conjecture has been inspirational, we have to change its statement for technical reasons and because we are not yet able to characterize non-minimal Hamiltonian systems. The main result of this monograph, theorem (5.11) may be stated as follows. : - If $\Phi_\Sigma(x_0)$ is the input-output map of an analytic, complete, system $\Sigma$, described by (1.1), which satisfies an additional assumption, then $\Phi_\Sigma(x_0)$ has a minimal, analytic, complete Hamiltonian realization $\Sigma'$, described by (1.15), if and only if for any $(\bar{u},\bar{y}) \in \Sigma_e^+(0)(x_0)$, any two admissible variations $(\delta_i u, \delta_i y)$, $i = 1, 2$, of $(\bar{u},\bar{y})$, such that $(\delta_i u, \delta_i y)$ have compact support in $(0, \infty)$, satisfy

$$\int_0^{\infty} (\delta_2^T y(t)\delta_1 u(t) - \delta_1^T y(t)\delta_2 u(t))dt = 0.$$

☐

The additional assumption is satisfied if for example $g_0(x_0) = 0$ in system (1.1), as assumed in theorem (1.3). The precise statement, as well as a definition of admissible, is given in section (5). If $\Sigma$ is minimal, then we may take $\Sigma' = \Sigma$, yielding theorem (5.9). This result is also true for general systems (1.2), see theorems (6.3) and (6.4).

We now describe a result more in the spirit of the original conjecture by Van der Schaft. It is necessary to consider certain infinite dimensional manifolds of maps $R \longrightarrow \Omega \times R^m$, $\Omega \subset R^m$, $t \longrightarrow (u(t),y(t))$, where $t \longrightarrow u(t)$ is piecewise constant and right continuous, and $t \longrightarrow y(t)$ is continuous. Even if both $u$ and $y$ were $C^\infty$ or $C^\omega$ it is not clear what topological of differentiable structure should be put on these "manifolds", since the domain of the functions is not compact. We shall therefore consider these only formally as manifolds, and derive formal results about then, which hopefully may be rigorized at a later date.

Consider first the manifold of maps $\mathbf{M}_{M,\Omega,m}$, defined as the union of all behaviour sets $\Sigma_e$ as $\Sigma$ ranges over all minimal, affine, analytic and complete systems (1.1) with state space M, control constraint set $\Omega \subset R^m$, and outputs in $R^m$, i.e. $m = p$. On this manifold we suppose the tangent space to it at $(u,y)$, $T_{(u,y)} \mathbf{M}_{M,\Omega,m}$, consists of all variational fields $(\delta u, \delta y)$, of compact support, corresponding to admissible variations of $(u,y)$. We define a (weak) symplectic form on $\mathbf{M}_{M,\Omega,m}$ as suggested by (1.33), by setting

$$(1.42) \qquad \mu_{(u,y)}\big((\delta_1 u, \delta_1 y),(\delta_2 u, \delta_2 y)\big) = \int_{-\infty}^{\infty} \big(\delta_2 y(t)^T \delta_1 u(t) - \delta_1 y(t)^T \delta_2 u(t)\big)dt$$

We now make the usual definitions. A submanifold $M \subset N_{M,\Omega,m}$ is isotropic if $\mu$ restricted to $M$ is identically zero. We say $M$ is a Lagrangian submanifold if it is isotropic and co-isotropic. To be precise, $M$ is co-isotropic if given $(u,y) \in M$, and $(Du,Dy) \in T_{(u,y)} N_{M,\Omega,m}$ then $\mu_{(u,y)}\big((\delta u, \delta y),(Du,Dy)\big) = 0$ for all $(\delta u, \delta y) \in T_{(u,y)}M$ implies that $(Du,Dy) \in T_{(u,y)}M$. Let $M_{\Sigma}^{H}$ be the submanifold of $N_{M,\Omega,m}$ consisting of the behaviour set $\Sigma_e$ of a Hamiltonian system $\Sigma$ described by equations (1.15). Our result, theorem (5.17), which is closest in spirit to that expressed in the conjecture of Van der Schaft may be stated as :

Every submanifold $M_{\Sigma}^{H}$ is a Lagrangian submanifold of $N_{M,\Omega,m}$ $\qquad\qquad$ □

These ideas can easily be extended to the systems (1.2). However we would like to present these results for the generalized systems introduced in chapter (6), as in the original conjecture, but we have not yet resolved all problems in dealing with such systems. Inparticular it is not clear how to deal with existence and uniqueness of minimal realizations when the external variables belong to a general manifold, as discussed earlier.

SECTION 1.7

It is clear from the statement of our main results, that our work does not solve the Hamiltonian realization problem as do the results of Jakubczyk in section (1.4); but rather they characterize those input-output maps, or external behaviours which have Hamiltonian realizations, as do theorems (1.2) and (1.3). However Jakubczyk's results are comprised of two parts, one part guarantees a realization, and the other part provides extra algebraic conditions which ensure that the realization may be taken to be Hamiltonian. One therefore naturally expects the extra algebraic conditions to be equivalent to the conditions we give. Indeed Van der Schaft [V2] shows the equivalence in the case of linear systems. The general situation will be discussed in chapter 7.

In chapter (2) of this monograph we introduce our version of the variational systems and adjoint variational systems, in the context of a control system (1.1). At a global level this involves the introduction of two new systems derived from (1.1), which we call the Hamiltonian extension, and the Prolongation. In chapter (3) we consider the minimality properties of both the Hamiltonian extension and Prolongation, as a result of minimality properties of the original system. In chapter (4) we introduce the concept of self adjointness for the variational systems, and establish an important intermediate result, theorem (4.2). This is simply stated as : - A minimal system is Hamiltonian if and only if it's variational systems are self adjoint. This result plays roughly the same role as theorem (1.5) in Jakubczyk's work [J1]. In chapter (5) the equivalence of self adjointness and the criterion (1.33) is established, along with a compilation of our results. In chapter (6) we outline the work required to generalize our results to systems described by equations (1.2). Finally in chapter (7) we show the equivalence of our self-adjointness condition with Jakubczyk's algebraic conditions (theorem 1.6), and we discuss some possible extensions to the theory presented in this monograph.

## 2. VARIATIONAL AND ADJOINT VARIATIONAL SYSTEMS

We are concerned with nonlinear control systems $\Sigma$ with an equal number of inputs $u_j$ and outputs $y_j$

$$(2.1) \quad \Sigma: \quad \begin{aligned} \dot{x} &= g_0(x) + \sum_{j=1}^{m} u_j g_j(x) \quad , \ x \in M \quad , \ x(0) = x_0 \\ y_j &= H_j(x) \quad , \ j = 1, \cdots, m \ , \ u = (u_1, \cdots, u_m) \in \Omega \subset \mathbb{R}^m \end{aligned}$$

As in section 1.1 of chapter (1) M denotes the _state space_, which is assumed to be a k-dimensional differentiable manifold, and $\Omega$ is the _control space_, which for simplicity is taken to be an open subset of $\mathbb{R}^m$, containing 0. The $u_j$ appearing in the right-hand side of the differential equation belong to a certain class of functions of time t, called the _admissible controls_. Basically a control function is admissible if the corresponding solution of the differential equation is defined. For our purposes we may restrict the admissible controls to the _piecewise constant right continuous_ functions. Moreover we assume the vectorfields $g_i$, $i = 0, 1, \cdots, m$, to be _complete_. This implies that for every control function the solution of the differential equation is well-defined for every $t \in \mathbb{R}$. Finally $H_j$, $j = 1, \cdots, m$, are functions from M to $\mathbb{R}$. Our major assumption will be that all data involved, i.e. M, $g_0, g_1, \cdots, g_m$, $H_1, \cdots, H_m$, are _real-analytic_.

A subclass of the nonlinear systems (2.1) is formed by the _Hamiltonian control systems_ as discussed in the Introduction and section 1.3 of chapter (1) see [V1] for a detailed treatment and references. Let M be a symplectic manifold with symplectic form $\omega$ (i.e. $\omega$ is a non-degenerate two-form such that $d\omega = 0$). Necessarily M is even-dimensional, say dim $M = k = 2n$. A vectorfield f on M is called _Hamiltonian_ if there exists a function $H : M \longrightarrow R$ such that $\omega(f, -) = - dH$. We then write $f = X_H$. Darboux's theorem ([A]) there exist coordinates $(q_1, \cdots, q_n, p_1, \cdots, p_n)$ for M such that locally $\omega = \sum_{i=1}^{n} dp_i \wedge dq_i$. Such coordinates are called _canonical_. In canonical coordinates a Hamiltonian vectorfield $X_H$ has the familiar form $\dot{q}_i = \dfrac{\partial H}{\partial p_i}$, $\dot{p}_i = - \dfrac{\partial H}{\partial q_i}$, $i = 1, \cdots, n$.

Now assume that the state space M in (2.1) is a symplectic manifold $(M, \omega)$ and that the vectorfields $g_j$ are given as $- X_{H_j}$, $j = 1, \cdots, m$. Furthermore suppose that _locally_ there exists a function $H_0$ such that $g_0 = X_{H_0}$. (This is equivalent to requiring that $L_{g_0} \omega = 0$, cf. [A]). $g_0$ is called a _locally_ Hamiltonian vectorfield and will be denoted by $X_0$. Then the resulting system

$$\dot{x} = X_0(x) - \sum_{j=1}^{m} u_j X_{H_j}(x) \quad , \quad x \in (M^{2n}, \omega), \quad x(0) = x_0$$

(2.2)

$$y_j = H_j(x) \quad j = 1, \cdots, m \quad , \quad u = (u_1, \cdots, u_m) \in \Omega \subset R^m$$

is called a <u>Hamiltonian system</u>. If there exists <u>globally</u> a function $H_0$ such that $g_0 = X_{H_0}$ then the system is called <u>globally</u> Hamiltonian.

We wish to give necessary and sufficient conditions for a <u>minimal</u> (as will be explained later on) nonlinear system (2.1) to be actually a Hamiltonian system (2.2), i.e. for the existence of a symplectic form on the state space M of (2.1) such that (2.1) equals (2.2). These conditions will be given entirely in terms of the input-output behavior of any <u>variational</u> system of (2.1) and the input-output behavior of a related linear system, called the <u>adjoint</u> system.

We shall now define the variational and adjoint system along a solution of a nonlinear system (2.1). For any initial state $x(0) = x_0$ we take a coordinate neighbourhood of M containing $x_0$ and let $x(t)$, $t \in [0, T]$, be the solution of (2.1) corresponding to an input function $u(t) = (u_1(t), \cdots, u_m(t))$ and the initial state $x(0) = x_0$ such that $x(t)$ remains within this coordinate neighborhood. Denote the resulting output by $y(t) = (y_1(t), \cdots, y_m(t))$ with $y_j(t) = H_j(x(t))$. Then the <u>variational system</u> along the state-input-output trajectory $(x(t), u(t), y(t))$ is given by the time-varying linear system

$$\dot{v}(t) = \frac{\partial g_0}{\partial x}(x(t))v(t) + \sum_{j=1}^{m} u_j(t) \frac{\partial g_j}{\partial x}(x(t))v(t) + \sum_{j=1}^{m} u_j^v g_j(x(t))$$

(2.3)

$$y_j^v(t) = \frac{\partial H_j}{\partial x}(x(t))v(t) \quad , \quad j = 1, \cdots, m \ , \ v(0) = v_0 \in R^k$$

where $\frac{\partial g_i}{\partial x}$ denotes the $k \times k$ Jacobian matrix of $g_i : R^k \longrightarrow R^k$ and $\frac{\partial H_j}{\partial x}$ is the $1 \times k$ Jacobian matrix of $H_j : R^k \longrightarrow R$. Furthermore $u^v = (u_1^v, \cdots, u_m^v) \in R^m$ and $y^v = (y_1^v, \cdots, y_m^v) \in R^m$ denote the inputs and outputs of the variational system. The system (2.5) is called variational because of the following.

Let $(x(t, \varepsilon), u(t, \varepsilon), y(t, \varepsilon))$, $t \in [a, b]$, be a <u>family</u> of state-input-output trajectories of (2.1), parametrized by $\varepsilon$, such that $x(t, 0) = x(t)$, $u(t, 0) = u(t)$ and $y(t, 0) = y(t)$, $t \in [a, b]$. Then the quantities

(2.4) $\qquad v(t) = \frac{\partial x(t, 0)}{\partial \varepsilon}$ , $u^v(t) = \frac{\partial u(t, 0)}{\partial \varepsilon}$ , $y^v(t) = \frac{\partial y(t, 0)}{\partial \varepsilon}$

satisfy (2.5). We note that in case of a fixed initial state $x(0) = x_0$ the variational state $v(0)$ at time 0 is necessarily 0.

The <u>adjoint (variational) system</u> along the same trajectory $(x(t), u(t), y(t))$ is obtained by "dualizing" the variational system to the linear time-varying system

$$(2.5) \quad -\dot{p}(t) = \left(\frac{\partial g_0}{\partial x}\right)^T \left(x(t)\right)p(t) + \sum_{j=1}^{m} u_j(t) \left(\frac{\partial g_j}{\partial x}\right)^T \left(x(t)\right)p(t) + \sum_{j=1}^{m} u_j^a \left(\frac{\partial H_j}{\partial x}\right)^T \left(x(t)\right)$$

$$y_j(t) = p^T(t) \, g_j\left(x(t\,)\right) , \; j = 1, \cdots, m \; , \quad p(0) = p_0 \in R^k$$

with inputs $u^a = (u_1^a, \cdots, u_m^a) \in R^m$ and outputs $y^a = (y_1^a, \cdots, y_m^a) \in R^m$ (T denotes transpose).

The fundamental lemma connecting variational and adjoint systems is

LEMMA 2.1
Along solutions of the variational and adjoint system corresponding to the same state-input-output trajectory the following identity holds

$$(2.6) \qquad \frac{d}{dt} \, p^T(t)v(t) = \left(u^v(t)\right)^T y^a(t) - \left(u^a(t)\right)^T y^v(t)$$

Furthermore the adjoint system is uniquely determined by (2.6).

Proof By direct differentiation we obtain

$$\frac{d}{dt} \, p^T(t)v(t) = \dot{p}^T(t)v(t) + p^T(t)\dot{v}(t) =$$

$$= \left(-p^T(t) \frac{\partial g_0}{\partial x} \left(x(t)\right) - \sum_{j=1}^{m} u_j(t) \, p^T(t) \frac{\partial g_j}{\partial x} \left(x(t)\right) - \sum_{j=1}^{m} u_j^a(t) \frac{\partial H_j}{\partial x} \left(x(t)\right)\right)v(t)$$

$$+ p^T(t)\left(\frac{\partial g_0}{\partial x}(x(t))v(t) + \sum_{j=1}^{m} u_j(t) \frac{\partial g_j}{\partial x} \left(x(t)\right)v(t) + \sum_{j=1}^{m} u_j^v(t)g_j\left(x(t)\right)\right) =$$

$$= - \sum_{j=1}^{m} u_j^a(t)y_j^v(t) + \sum_{j=1}^{m} u_j^v(t)y_j^a(t).$$

Now let $\dot{p}(t) = F(t)p(t) + G(t) \, u^a(t)$, $y^a(t) = H(t)p(t)$ be an arbitrary time-varying linear system. Suppose it satisfies (2.6) for any $u^v(t)$, $u^a(t)$. Then necessarily $F(t) = - \left(\frac{\partial g_0}{\partial x}\right)^T(x(t)) - \sum_{j=1}^{m} u_j(t)\left(\frac{\partial g_j}{\partial x}\right)^T(x(t))$, the j-th column of $G(t)$ equals $- \left(\frac{\partial H_j}{\partial x}\right)^T (x(t))$ and the j-th row of $H(t)$ is $g_j^T(x(t))$. So the system equals the adjoint system. □

We may also add the variational or adjoint system to the original system (2.1) and regard them as one system. We call the original system together with the variational system, i.e.,

$$\dot{x}(t) = g_0(x(t)) + \sum_{j=1}^{m} u_j(t)g_j(x(t))$$

$$\dot{v}(t) = \frac{\partial g_0}{\partial x}(x(t))v(t) + \sum_{j=1}^{m} u_j(t) \frac{\partial g_j}{\partial x}(x(t))v(t) + \sum_{j=1}^{m} u_j^v(t)g_j(x(t))$$

(2.7)

$$y_j(t) = H_j(x(t))$$

$$j = 1, \cdots, m$$

$$y_j^v(t) = \frac{\partial H_j}{\partial x}(x(t))v(t)$$

with inputs $u_j$ and $u_j^v$, outputs $y_j$ and $y_j^v$ and state $(x,v)$, the _prolongation_ of (2.1) or _prolonged system_. The original system together with the adjoint system

$$\dot{x}(t) = g_0(x(t)) + \sum_{j=1}^{m} u_j(t)g_j(x(t))$$

$$\dot{p}(t) = -\left(\frac{\partial g_0}{\partial x}\right)^T (x(t))p(t) - \sum_{j=1}^{m} u_j(t)\left(\frac{\partial g_j}{\partial x}\right)^T (x(t))p(t)$$

(2.8)

$$- \sum_{j=1}^{m} u_j^a(t)\left(\frac{\partial H_j}{\partial x}\right)^T (x(t))$$

$$y_j(t) = H_j(x(t))$$

$$j = 1, \cdots, m$$

$$y_j^a(t) = p^T(t)g_j(x(t))$$

with inputs $(u, u^a)$, output $(y, y^a)$ and state $(x,p)$, is called the _Hamiltonian_ _extension_ of (2.1). This terminology will become clear as we will now give a _coordinate-free_ definition of both systems (2.7) and (2.8), which also shows that the prolongation and Hamiltonian extension are _globally_ (not just in a coordinate neighbourhood) defined systems.

First we give the definition of a _prolongation_ (or complete lift, cf. [Y]) of a function and a vectorfield. Let $H : M \longrightarrow R$, then the prolongation $\dot{H} : TM \longrightarrow R$ is defined by

(2.9)     $\dot{H}(x,v) = dH(x)v$  ,   $v \in T_x M$

Given local coordinates $(x_1, \cdots, x_k)$ for M we obtain natural coordinates $(x_1, \cdots, x_k, v_1 = \dot{x}_1, \cdots, v_k = \dot{x}_k)$ for TM. In these coordinates, H is just given by

(2.10)     $\dot{H}(x,v) = \sum_{i=1}^{k} \frac{\partial H}{\partial x_j}(x)v_j .$

Let f be a vectorfield on M, with integral flow $f_t : M \longrightarrow M$, $t \in [0,\varepsilon)$. Then $(f_t)_* : TM \longrightarrow TM$ is the integral flow of the prolonged vectorfield $\dot{f}$ on TM. In the above natural coordinates

$$(2.11) \qquad \dot{f}(x,v) = \sum_{i=1}^{k} f_i(x) \frac{\partial}{\partial x_i} + \sum_{i,j=1}^{k} \frac{\partial f_i}{\partial x_j}(x) \, v_j \frac{\partial}{\partial v_j}$$

Denote the natural projection from TM to M by $\pi$. Then for any function $H : M \longrightarrow R$ we define the <u>vertical lift</u> (cf. [Y]) $H^{\ell} : TM \longrightarrow R$ of H simply by

$$(2.12) \qquad H^{\ell} = H \circ \pi$$

In local coordinates $H^{\ell}(x,v) = H(x)$. For any vectorfield $f$ on M we let the vertical lift $f^{\ell}$ be the vectorfield on TM such that

$$(2.13) \qquad f^{\ell}(\dot{H}) = \left( f(H) \right)^{\ell} \quad \text{for any} \quad H : M \longrightarrow R$$

In [Y] it is shown that this determines $f^{\ell}$ uniquely as a vectorfield, and moreover that $f^{\ell}$ in natural coordinates is simply given as

$$(2.14) \qquad f^{\ell}(x,v) = \sum_{i=1}^{k} f_i^{\ell}(x) \frac{\partial}{\partial v_i}$$

After these preparations, we define the prolongation of (2.1) as the system

$$\dot{x}_p = \dot{g}_0(x_p) + \sum_{j=1}^{m} u_j \dot{g}_j(x_p) + \sum_{j=1}^{m} u_j^v g_j^{\ell}(x_p)$$

$$(2.15) \qquad y_j = H_j^{\ell}(x_p) \qquad\qquad\qquad x_p \in TM, \ x_p(0) = (x_0, v_0)$$

$$\qquad\qquad\qquad\qquad j = 1, \cdots, m$$

$$y_j^v = \dot{H}_j(x_p) \qquad\qquad\qquad u = (u_1, \cdots, u_m) \in \Omega \subset R^m$$

$$\qquad\qquad\qquad\qquad\qquad\qquad u^v = (u_1^v, \cdots, u_m^v) \in R^m$$

It is easily seen that in natural coordinates $(x,v)$ for TM (2.15) reduces to (2.7). For the definition of the Hamiltonian extension we note that $T^*M$ as a cotangent bundle has a canonically defined symplectic form $\Omega$. In natural coordinates $(x_1, \cdots, x_k, p_1, \cdots, p_k)$ for $T^*M$, $\Omega$ is given by $\sum_{i=1}^{k} dp_i \wedge dx_i$. Furthermore with any vectorfield $f$ on M we associate a function $H^f$ from $T^*M$ to R by setting

$$(2.16) \qquad H^f(x,p) = \langle p, f(x) \rangle = p^T f(x) , \quad p \in T_x^* M$$

where $\langle , \rangle$ is the natural pairing between $T_x M$ and $T_x^* M$. For notational ease we will often write $p^T f(x)$ instead of $H^f$. Finally denote the projection from $T^*M$ to M by $\pi$. Then the vertical lift of a function H on M is again defined by

(2.17)     $H^\ell = H \circ \pi$.

The Hamiltonian extension of (2.1) is now given as

$$\dot{x}_e = X_{H^{g_0}}(x_e) + \sum_{j=1}^{m} u_j X_{H^{g_j}}(x_e) + \sum_{j=1}^{m} u_j^a X_{H_j^\ell}(x_e)$$

(2.18)

$$y_j = H_j^\ell(x_e) \qquad\qquad x_e \in T^*M, \ x_e(0) = (x_0, p_0)$$

$$\qquad\qquad j = 1, \cdots, m$$

$$y_j^a = H^{g_j}(x_e) \qquad\qquad u = (u_1, \cdots, u_m) \in \Omega \subset \mathbb{R}^m$$

$$\qquad\qquad u^a = (u_1^a, \cdots, u_m^a) \in \mathbb{R}^m$$

where of course $X_H$ denotes the Hamiltonian vectorfield on $T^*M$ defined by $\Omega(X_H, -) = -dH$, $H : T^*M \longrightarrow \mathbb{R}$. It is easily seen that in natural coordinates $(x,p)$ for $T^*M$ (2.18) reduces to the expression (2.8) given before. We note that the Hamiltonian extention is itself a globally <u>Hamiltonian</u> system. (In order to keep the same sign convention as in (2.2) one should take $-u^a$ instead of $u^a$ and $-y^a$ instead of $y^a$. Notice furthermore that the inputs $u_j$ correspond to the "adjoint" outputs $y_j^a$, and the "adjoint" inputs $u_j^a$ correspond to the outputs $y_j$, see also chapter (6)).

We conclude that the prolongation and Hamiltonian extension are globally defined systems on $TM$, resp. $T^*M$. In contrast, the variational system (2.3) and adjoint system (2.5) have been only defined on coordinate neighbourhoods of $M$. It is easily seen that these definitions actually can be extended to <u>trivializing charts</u> of $TM$, resp. $T^*M$. However in general (except for the case that $TM$ and $T^*M$ are trivial bundles) the variational and adjoint system ca<u>nnot</u> be globally defined.

The above coordinate-free definitions of prolongation and Hamiltonian extension also enable us to give a coordinate-free version of Lemma 2.1. Namely

$$(2.19) \qquad \frac{d}{dt} \langle p(t), v(t) \rangle_{x(t)} = \begin{bmatrix} y^v(t) \\ u^v(t) \end{bmatrix}^T J^e \begin{bmatrix} y^a(t) \\ u^a(t) \end{bmatrix}$$

where $J^e$ is the linear symplectic form $\begin{bmatrix} 0 & -I_m \\ I_m & 0 \end{bmatrix}$ on $\mathbb{R}^m \times \mathbb{R}^m$, $\frac{d}{dt}$ denotes differentiation along the prolongation and Hamiltonian extension and $\langle,\rangle_x$ denotes the pairing between $T_xM$ and $T_x^*M$.

## 3. MINIMALITY OF THE PROLONGATION AND HAMILTONIAN EXTENSION

Consider a nonlinear system (2.1). Denote by $L$ the Lie algebra generated by all the vectorfields $g_0, g_1, \ldots, g_m$ under Lie bracketing. Denote by $L_0$ the ideal in $L$ generated by the vectorfields $g_1, \ldots, g_m$. $L_0$ is called the accessibility algebra and $L_0$ the strong accessibility algebra. It is well-known [S2] that the system is strongly accessible if and only if dim $L_0(x)$ = dim M for any $x \in M$, where $L_0(x)$ = = $\text{span}_R \{f(x) | f \in L_0\} \subset T_x M$. (Strong accessibility means that the set of points which can be reached at any time T > 0 from any point $x \in M$ by choosing different input functions contains a non-empty interior with respect to M ([S2]).)

Furthermore denote by $H$ the linear space of functions of the form $L_{f_1} L_{f_2} \ldots L_{f_s} H_j$, with $f_r$, $r = 1, \ldots, s$, equal to $g_i$, $i = 0, 1, \ldots, m$, and $j = 1, \ldots, m$. Equivalently ([H]), we may take $f_r$ to be arbitrary elements of $L$. $H$ is called the observation space. It follows from the analyticity assumption that the system is observable if and only if $H$ distinguishes points in M, i.e. for every $x_1, x_2 \in M$ with $x_1 \neq x_2$ there exists an $H \in H$ such that $H(x_1) \neq H(x_2)$, cf. [H]. (Observability means that for every pair $x_1, x_2 \in M$ with $x_1 \neq x_2$ there exists an admissible control such that the output functions resulting from the initial conditions x(0) = $x_1$, resp. x(0) = $x_2$, are different). The system is weakly observable if $H$ only distinguishes nearby points in M.

We call a nonlinear system minimal if it is strongly accessible as well as observable. This definition is slightly stronger than the usual one where one only requires observability and accessibility, i.e. dim L(x) = dim M for any $x \in M$, cf. [H]. (The reason is that for Hamiltonian systems observability and accessibility implies strong accessibility; see Remark 2 after Proposition 3.4.) A nonlinear system is called quasi minimal if it is strongly accessible and weakly observable. We will now relate the accessibility and the observability properties of a nonlinear system (2.1) and its prolongation. The basic connection is contained in

## THEOREM 3.1

Consider a nonlinear system $\Sigma$ given by (2.1) with strong accessibility algebra $L_0$ and observation space $H$.

a.     The strong accessibility algebra of the prolongation is given by $L_0^p = \dot{L}_0 + L_0^{\ell}$, where $L_0^{\ell} = \{f^{\ell} | f \in L_0\}$ and $\dot{L}_0 = \{\dot{f} | f \in L_0\}$.

b.     The observation space of the prolongation is given by $H^p = \dot{H} + H^{\ell}$, where $H^{\ell} = \{H^{\ell} | H \in H\}$ and $\dot{H} = \{\dot{H} | H \in H\}$.

In order to facilitate the proof of this theorem and for later use we first list some identities for Lie derivatives of prolongations and vertical lifts of vectorfields and functions.

## LEMMA 3.2

Let M be a manifold. Then for any vectorfields $f$, $f_1, f_2$ on M and any function $H : M \to R$ the following identities hold:

a. $\quad [\dot{f}_1, \dot{f}_2] = \overline{[f_1, f_2]}\,\dot{}$

b. $\quad [\dot{f}_1, f_2^{\ell}] = [f_1, f_2]^{\ell}$

c. $\quad [f_1^{\ell}, f_2^{\ell}] = 0$

d. $\quad \dot{f}(\dot{H}) = \overline{f(H)}\,\dot{}$

e. $\quad \dot{f}(H^{\ell}) = (f(H))^{\ell} = f^{\ell}(\dot{H})$

f. $\quad f^{\ell}(H^{\ell}) = 0$

Proof a. In natural coordinates $(x,v)$ for TM

$$\dot{f}_j = f_j \frac{\partial}{\partial x} + \frac{\partial f_j}{\partial x} v \frac{\partial}{\partial v} , \; j = 1,2. \quad \text{Hence}$$

$$[\dot{f}_1, \dot{f}_2] = \begin{bmatrix} \frac{\partial f_2}{\partial x} & 0 \\ \frac{\partial}{\partial x}\left(\frac{\partial f_2}{\partial x} v\right) & \frac{\partial f_2}{\partial x} \end{bmatrix} \begin{bmatrix} f_1 \\ \frac{\partial f_1}{\partial x} v \end{bmatrix} - \begin{bmatrix} \frac{\partial f_1}{\partial x} & 0 \\ \frac{\partial}{\partial x}\left(\frac{\partial f_1}{\partial x} v\right) & \frac{\partial f_1}{\partial x} \end{bmatrix} \begin{bmatrix} f_2 \\ \frac{\partial f_2}{\partial x} v \end{bmatrix} =$$

$$\begin{bmatrix} \frac{\partial f_2}{\partial x} f_1 - \frac{\partial f_1}{\partial x} f_2 \\ \frac{\partial}{\partial x}\left(\frac{\partial f_2}{\partial x} v\right) f_1 + \frac{\partial f_2}{\partial x}\frac{\partial f_1}{\partial x} v - \frac{\partial}{\partial x}\left(\frac{\partial f_1}{\partial x} v\right) f_2 - \frac{\partial f_1}{\partial x}\frac{\partial f_2}{\partial x} v \end{bmatrix} = \begin{bmatrix} [f_1, f_2] \\ \frac{\partial}{\partial x} [f_1, f_2] v \end{bmatrix} =$$

$$\overline{[f_1, f_2]}\,\dot{}$$

b. Since $f_2^\ell = f_2 \frac{\partial}{\partial v}$ we have

$$[\dot{f}_1, f_2^\ell] = \begin{bmatrix} 0 & 0 \\ \frac{\partial f_2}{\partial x} & 0 \end{bmatrix} \begin{bmatrix} f_1 \\ \frac{\partial f_1}{\partial x} v \end{bmatrix} - \begin{bmatrix} \frac{\partial f_1}{\partial x} & 0 \\ \frac{\partial}{\partial x}(\frac{\partial f_1}{\partial x} v) & \frac{\partial f_1}{\partial x} \end{bmatrix} \begin{bmatrix} 0 \\ f_2 \end{bmatrix} =$$

$$(\frac{\partial f_2}{\partial x} f_1 - \frac{\partial f_1}{\partial x} f_2) \frac{\partial}{\partial v} = [f_1, f_2]^\ell$$

c. can be proved similary, while d,e,f can be proved using the local coordinate expressions for $\dot{H}$ and $H^\ell$. □

Proof of Theorem 3.1

a. The accessibility algebra $L^p$ of the prolongation is generated by the vectorfields $\dot{g}_0, \dot{g}_j, g_j$, $j = 1, \ldots m$, on TM. Since by lemma 3.2 a,b,c,

$$[\dot{g}_1, \dot{g}_j^\ell] = \overline{[\dot{g}_1, g_j]}, \quad [\dot{g}_1, g_j^\ell] = [g_1, g_j]^\ell \text{ and } [g_1^\ell, g_j^\ell] = 0, \quad i = 0,1,\ldots,m, j = 1,\ldots,m,$$

we conclude that $L^p = \dot{L} + L^\ell$, where of course $\dot{L} = \{\dot{f} | f \in L\}$ and $L^\ell = \{f^\ell | f \in L\}$. It follows that the ideal $L_0^p \subset L^p$ generated by $\dot{g}_j, g_j^\ell$, $j = 1, \ldots, m$ is given as $L_0^p = \dot{L}_0 + L_0^\ell$.

b. The observation space $H^p$ of the prolongation is spanned by all functions $L_{k_1} L_{k_2} \cdots L_{k_s} H_j^\ell$ and $L_{k_1} L_{k_2} \cdots L_{k_s} \dot{H}_j$, with $k_r$, $r = 1,\ldots,s$, equal to $\dot{g}_1$, $i = 0,1,\ldots,m$, or $g_j^\ell$, $j = 1,\ldots,m$. Using Lemma 3.2 d, e, f it follows that $H^p$ is spanned by the functions $(L_{f_1} L_{f_2} \cdots L_{f_s} H_j)^\ell$ and $L_{f_1} L_{f_2} \cdots L_{f_s} H_j$, with $f_r$, $r = 1,\ldots,s$ equal to $g_1$, $i = 0,1,\ldots,m$. Hence $H^p = \dot{H} + H^\ell$. □

The main use of Theorem 3.1 is contained in

COROLLARY 3.3

Consider a nonlinear system $\Sigma$ given by (2.1). Then

a. $\Sigma$ is strongly accessible if and only if its prolongation is strongly accessible.

b. Let $\Sigma$ be strongly accessible. Then $\Sigma$ is (weakly) observable if and only if its prolongation is (weakly) observable.

c. $\Sigma$ is (quasi-)minimal if and only if its prolongation is (quasi-) minimal.

Proof a. By Theorem 3.1 $L_0^P = \dot{L}_0 + L_0^\ell$. Hence $\dim L_0^P(x_p) = \dim TM$ for any $x_p \in$ TM, if and only if $\dim L_0(x) = \dim M$ for every $x \in M$.

b. By Theorem 3.1 $H^P = \dot{H} + H^\ell$. Since $\Sigma$ is (strongly) accessible and analytic $dH(x):= \text{span}_R\{dH(x)|H \in \mathbf{H}\}$ has constant dimension ([H]). Similarly $dH^P(x_p)$ has constant dimension for every $x_p \in$ TM. Furthermore $\Sigma$ is weakly observable if and only $\dim dH(x) = \dim M([H])$ and the prolongation is weakly observable if and only if $\dim d\mathbf{H}^P = \dim TM$. Hence $\Sigma$ is weakly observable if and only if its prolongation is weakly observable. Now suppose the prolongation is observable. Take two points $x_1, x_2 \in M$, with $x_1 \neq x_2$. Then $(x_1,0)$ and $(x_2,0) \in$ TM. Hence there exists a function $H^P \in \mathbf{H}^P$ such that $H^P(x_1,0) \neq H^P(x_2,0)$. Since $\dot{H}(x,0) = 0$ for every $x \in M$ and $H \in \mathbf{H}$ it follows that there exists a function $H \in \mathbf{H}$ such that $H^\ell(x_1,0) \neq H^\ell(x_2,0)$ or equivalently $H(x_1) \neq H(x_2)$. Hence $\Sigma$ is observable. Conversely assume that $\Sigma$ is observable. Take two points $(x_1,v_1)$ and $(x_2,v_2)$ in TM. Suppose $H^P(x_1,v_1) = H^P(x_2,v_2)$ for every $H^P \in \mathbf{H}^P$. Since $\mathbf{H}^\ell \subset \mathbf{H}^P$ this yields $x_1 = x_2 = x$. Then, since $\dot{\mathbf{H}} \subset \mathbf{H}^P$ we have that $\frac{\partial H}{\partial x}(x)v_1 = \frac{\partial H}{\partial x}(x)v_2$ for all $H \in \mathbf{H}$. Because $\dim dH(x) = \dim M$ for any $x \in M$, we obtain $v_1 = v_2$.

c. Follows immediately from a. and c.                                       □

For the connections between strong accessibility and observability of a nonlinear system and of its Hamiltonian extension, it appears that we have to combine both properties. This is due to the fact that for a Hamiltonian system (and hence for the Hamiltonian extension) strong accessibility is almost equivalent with observability. Recall that on a symplectic manifold $(M,\omega)$ the Poisson bracket of two functions is defined as

$$(3.1) \qquad \{F,G\} = \omega(X_F, X_G) = X_F(G), \qquad F,G: M \rightarrow R,$$

and satisfies the basic identity

$$(3.2) \qquad [X_F, X_G] = X_{\{F,G\}}$$

This makes $C^\infty(M)$ into a Lie algebra under Poisson brackets and by (3.2) $C^\infty(M)$ modulo the constant functions is isomorphic to the Lie algebra (under Lie brackets) of Hamiltonian vectorfields on $(M,\omega)$. In canonical coordinates $(q_1,...,q_n, p_1,...,p_n)$ the Poisson bracket equals the familiar expression

$$(3.3) \qquad \{F,G\} = \sum_{i=1}^{n} \left( \frac{\partial F}{\partial p_i} \frac{\partial G}{\partial q_i} - \frac{\partial F}{\partial q_i} \frac{\partial G}{\partial p_i} \right)$$

PROPOSITION 3.4 ([V1,V4])

Consider a Hamiltonian system (2.2). The observation space $\mathbf{H}$ is equal to the ideal generated by the functions $H_1,...,H_m$ within the Lie algebra (under Poisson brackets) generated by the functions $H_0$, $H_1,...,H_m$.
Furthermore the strong accessibility algebra $\mathbf{L}_0$ is equal to the set of vectorfields $X_H$ with $H \in \mathbf{H}$. Consequently $\dim \mathbf{L}_0(x) = \dim M \Longleftrightarrow \dim d\mathbf{H}(x) = \dim M$
$(d\mathbf{H}(x) = \text{span}_R\{d\mathbf{H}(x)|H \in \mathbf{H}\})$.

Proof ([V1,V4]) $\mathbf{H}$ is generated by functions $L_{f_1} L_{f_2} .. L_{f_s} H_j$ with $j=1, .., m$, and $f_r$, $r=1,..,s$, equal to $X_{H_i}$, $i=0,1,..,m$. By (3.1) this is equal to
$\{F_1,\{F_2,\{...\{F_s,H_j\}...\}\}\}$, with $j=1,..,m$, and $F_r$, $r=1,...,s$, equal to $H_i$, $i=0,1,$
$...,m$. Furthermore $\mathbf{L}_0$ is generated by vectorfields $[f_1,[f_2,[.... [f_s, X_{H_j}]..]]]$
with $j=1,.. , m$, and $f_r$, $r=1,...,s$, equal to $X_{H_i}$, $i=0,1,...,m$. By (3.2) this is
equal to a Hamiltonian vectorfield corresponding to a function in $\mathbf{H}$. $\qquad \Box$

Remark 1 Note that even if $H_0$ is locally defined, the observation space $\mathbf{H}$ consists of globally defined functions.

Remark 2 For an accessible analytic system the dimension of $d\mathbf{H}(x)$ is constant ([H]). Hence it follows from Proposition 3.4 that an accessible, weakly observable Hamiltonian system is automatically strongly accessible.

THEOREM 3.5

Consider a nonlinear system $\Sigma$ given by (2.1) with strong accessibility algebra $\mathbf{L}_0$ and observation space $\mathbf{H}$. Then the observation space of the Hamiltonian extension is given by $\mathbf{H}^e = p^T\mathbf{L}_0 + \mathbf{H}^\ell$, where $p^T\mathbf{L}_0 = \{p^Tf(x)|f \in \mathbf{L}_0\}$ and $\mathbf{H}^\ell = \{H^\ell|H \in \mathbf{H}\}$.

For the proof we again need some identities.

LEMMA 3.6.

Let M be a k-dimensional manifold. Consider T*M with its natural symplectic form $\Omega$, in natural coordinates $(x_1,\ldots,x_k,p_1,\ldots,p_k)$ given by $\Omega = \sum\limits_{i=1}^{k} dp_i \wedge dx_i$, and the corresponding Poisson bracket

$$(3.4) \qquad \{F,G\} = \sum_{i=1}^{k} \left( \frac{\partial F}{\partial p_i} \frac{\partial G}{\partial x_i} - \frac{\partial F}{\partial x_i} \frac{\partial G}{\partial p_i} \right)$$

Then for any vectorfields $f,f_1,f_2$ on M and functions $H,H_1,H_2$ on M we have

a. $\qquad X_{p^T f_1(x)}\left(p^T f_2(x)\right) = \{p^T f_1(x),\ p^T f_2(x)\} = p^T[f_1,f_2](x)$

b. $\qquad X_{p^T f(x)}(H^\ell) = \{p^T f(x),H^\ell\} = (f(H))^\ell = -X_{H^\ell}(p^T f(x))$

c. $\qquad X_{H_1^\ell}(H_2^\ell) = \{H_1^\ell,H_2^\ell\} = 0$

d. $\qquad [X_{p^T f_1(x)},\ X_{p^T f_2(x)}] = X_{\{p^T f_1(x),p^T f_2(x)\}} \overset{(a)}{=} X_{p^T[f_1,f_2](x)}$

e. $\qquad [X_{p^T f(x)},\ X_{H^\ell}] = X_{\{p^T f(x),H^\ell\}} \overset{(b)}{=} X_{(f(H))^\ell}$

f. $\qquad [X_{H_1^\ell},\ X_{H_2^\ell}] = X_{\{H_1^\ell,H_2^\ell\}} \overset{(c)}{=} 0$

Proof: Use the local coordinate expression (3.4) for the Poisson bracket. $\qquad \square$

Proof of Theorem 3.5

$H^e$ is spanned by all functions of the form $\{G_1,\{G_2,\{G_3,\ \ldots\{G_s,G\}\ldots\}\}\}$ with $G_r$, $r=1,\ldots,s$, equal to a function $p^T g_i(x)$, $i=0,1,\ldots,m$, or $H_j^\ell$, $j=1,\ldots,m$, and G equal to a function $p^T g_j(x)$ or $H_j^\ell$, $j=1,\ldots,m$. Therefore using Lemma 3.6, $H^e$ is spanned by all functions

$$p^T[f_1,[f_2,[f_3,\ldots,[f_s,f]..]]]$$

with $f_r$, $r=1,\ldots,s$, equal to a vectorfield $g_i$, $i=0,1,\ldots,m$, and f equal to a vectorfield $g_j$, $j=1,\ldots,m$, and all functions

$$\{p^T f_1(x), \{p^T f_2(x), \{p^T f_3(x) \ldots \{p^T f_s(x), H_j^{\ell}(x_e)\} \ldots \}\}\} =$$

$$= (L_{f_1} L_{f_2} L_{f_3} \ldots L_{f_s} H_j)^{\ell} \qquad j=1, \ldots, m,$$

with $f_r$, $r=1, \ldots, s$ equal to $g_i$, $i=0, 1, \ldots, m$. Hence $H^e = p^T L_0 + H^{\ell}$. $\qquad \square$

## COROLLARY 3.7

Consider a nonlinear system $\Sigma$ given by (2.1). Then $\Sigma$ is (quasi-) minimal if and only if its Hamiltonian extension is (quasi-) minimal.

Proof Let $\Sigma$ be minimal. Take $(x_1, p_1)$ and $(x_2, p_2)$ in $T^*M$ such that $H^e(x_1, p_1) = H^e(x_2, p_2)$ for all $H^e \in \mathbf{H}^e$. Since $H^{\ell} \subset H^e$ and $\Sigma$ is observable this implies $x_1 = x_2 = x$. Furthermore since $p^T L_0 \subset H^e$ we then have $p_1^T f(x) = p_2^T f(x)$ for every $f \in L_0$. Because $\Sigma$ is strongly accessible this yields $p_1 = p_2$. Hence the Hamiltonian extension is observable, and by Proposition 3.4 strongly accessible. Conversely assume the Hamiltonian extension to be minimal. Since the prolongation is real-analytic and $H^e = p^T L_0 + H^{\ell}$ this implies that $\dim d\mathbf{H}^e(x_e) = \dim T^*M$ and $\dim L_0(x) = \dim M$ for every $x_e \in T^*M$, $x \in M$. Furthermore because $\mathbf{H}^{\ell} \subset \mathbf{H}^e$ the system $\Sigma$ is observable. $\qquad \square$

Remark Note that in general the relation between the accessibility and observability properties of a nonlinear system and its variational or adjoint systems (both viewed as time-varying linear systems) is not so clear. On the other hand it can be easily proven that a variational system is controllable, resp. observable, if and only if the corresponding adjoint system is observable, resp. controllable.

For later use we remark that regarding the observability properties of the prolongation, resp. Hamiltonian extension, we actually do not need the additional inputs $u^v$, resp. $u^a$, to distinguish between two states. Consequently we may put $u^v$ and $u^a$ equal to zero:

<u>PROPOSITION 3.8</u>

Consider the system (2.1) with observation space $\mathbf{H}$ and strong accessibility algebra $\mathbf{L}_0$.

a. The observation space of the truncated prolongation

$$\dot{x}_p = \dot{g}_0(x_p) + \sum_{j=1}^{m} u_j \, \dot{g}(x_p) \qquad x_p \in TM$$

(3.4)
$$y_j = H_j^{\ell}(x_p)$$
$$\qquad\qquad j=1,\ldots,m$$
$$y_j^{\,v} = \dot{H}_j(x_p)$$

is equal to the observation space of the prolongation, i.e. $\dot{H} + H^{\ell}$. Hence if (2.1) is (quasi-) minimal, then (3.4) is (weakly) observable.

b. The observation space of the truncated Hamiltonian extension

$$\dot{x}_e = X_{H^{g_0}}(x_e) + \sum_{j=1}^{m} u_j \, X_{H^{g_j}}(x_e) \qquad x_e \in T^*M$$

(3.5)
$$y_j = H_j^{\ell}(x_e)$$
$$\qquad\qquad j=1,\ldots,m$$
$$y_j^{\,a} = H^{g_j}(x_e)$$

equals the observation space of the Hamiltonian extension, i.e. $p^T \mathbf{L}_0 + H^{\ell}$. Hence if (2.1) is (quasi-) minimal, then (3.5) is (weakly) observable.

<u>Proof</u> Consider the proof of Theorem 3.1 a and note that the terms $f^{\ell}(H^{\ell})$ and $f^{\ell}(\dot{H})$, with $f \in \mathbf{L}_0$ and $h \in \mathbf{H}$, are superfluous in the construction of $\mathbf{H}^p$. Hence the observation space of (3.4) equals $H^p = \dot{H} + H^{\ell}$. Similarly, observe in the proof of theorem 3.5 that for the construction of $H^e = p^T \mathbf{L}_0 + H^{\ell}$ the terms $X_{H^{\ell}}(\mathbf{H})$ and $X_{H^{\ell}}(p^T \mathbf{L}_0)$, with $H \in \mathbf{H}$, are redundant. Observability follows as in Corollary 3.3, resp. Corollary 3.7. $\qquad\qquad\qquad\square$

## 4. THE SELF-ADJOINTNESS CRITERION

In this chapter we shall show that a minimal nonlinear system (2.1) is a Hamiltonian system (2.2) if and only if every variational system is self-adjoint, which will be interpreted as meaning that the input-output maps of the variational system and the adjoint system coincide.

So let us consider the analytic nonlinear system (2.1)

(4.1)
$$\dot{x} = g_0(x) + \sum_{j=1}^{m} u_j\, g_j(x) \ , \ x \in M, \qquad x(0) = x_0$$

$$y_j = H_j(x), \qquad j=1,\ldots,m, \qquad\qquad u \in \Omega \subset R^m$$

in which all the associated vectorfields $g_0 + \sum_j u_j g_j$, $u \in \Omega$, are complete. In any local coordinate chart $(x,v)$ for TM the variational system along an admissible control function $u(t)$ (e.g. piecewise constant right continuous) is given as

(4.2)
$$\dot{v}(t) = A(t)v(t) + B(t)u^v(t), \qquad\qquad v(0) = v_0$$

$$y^v(t) = C(t)v(t)$$

and in a local coordinate chart $(x,p)$ for T*M the adjoint system is given as

(4.3)
$$\dot{p}(t) = -A^T(t)p(t) - C^T(t)u^a(t), \qquad\qquad p(0) = p_0$$

$$y^a(t) = B^T(t)p(t)$$

where

(4.4)
$$A(t) = \frac{\partial g_0}{\partial x}(x(t)) + \sum_{j=1}^{m} u_j(t) \frac{\partial g_j}{\partial x}(x(t))$$

$$B(t) = (g_1(x(t)) \,|\, \ldots \,|\, g_m(x(t)))$$

$$C(t) = (\frac{\partial H_j}{\partial x_i}(x(t)))\ \begin{matrix} j=1,\ldots,m \\ i=1,\ldots,k \end{matrix}$$

In such local coordinates the input-output map of the variational system for $v_0 = 0$ is given by

(4.5) $\qquad y^v(t) = \int_0^t W_v(t,\sigma,u) \, u^v(\sigma)d\sigma, \qquad 0 \le t \le \tau$

Where $W_v(t,\sigma,u) = C(t)\phi^u(t,\sigma)B(\sigma)$, with $\phi^u(t,\sigma)$ the unique solution of

(4.6) $\qquad \dfrac{\partial}{\partial t} \, \phi^u(t,\sigma) = A(t)\phi^u(t,\sigma), \quad \phi^u(\sigma,\sigma) = I_k$

Similarly, locally the input-output map of the adjoint system for $p_0 = 0$ is given by

(4.7) $\qquad y^a(t) = \int_0^t W_a(t,\sigma,u)u^a(\sigma)d\sigma, \qquad 0 \le t \le \tau$

with $W_a(t,\sigma,u) = -B^T(t)\psi^u(t,\sigma)C^T(\sigma)$ and $\psi^u(t,\sigma)$ the unique solution of

(4.8) $\qquad \dfrac{\partial}{\partial t} \, \psi^u(t,\sigma) = -A^T(t)\psi^u(t,\sigma), \quad \psi(\sigma,\sigma) = I_k$

As a matter of fact it follows from $\phi^u(t,\sigma)\phi^u(\sigma,t) = \phi^u(t,t)$ and (4.6) that $(\phi^u(\sigma,t))^T$ satisfies (4.8), and so actually

$$W_a(t,\sigma,u) = -B^T(t)\big(\phi^u(\sigma,t)\big)^T C^T(\sigma) = -W_v^T(\sigma,t,u).$$

We shall now show how $W_v(t,\sigma,u)$ and $W_a(t,\sigma,u)$ can be defined for all $t$, $\sigma \ge 0$.

Let $(t,\sigma,x) \to \psi^u_{t,\sigma}(x)$ denote the flow of the time-varying vectorfield $g_0(x) + \sum_{j=1}^m u_j(t)g_j(x)$, where $u_j(t), t \in [0,\infty)$, $j=1,\ldots, m$, are piecewise constant. Because by assumption the vectorfields $g_0(x) + \sum_{j=1}^m u_j g_j(x)$ are for every $(u_1,\ldots,u_m) \in \Omega$ __complete__, $\psi^u$ is defined for all $t \ge 0$, $\sigma \ge 0$ and $x \in M$. Indeed each mapping $x \to \psi^u_{t,\sigma}(x)$ is the concatenation of a finite number of diffeomorphisms $x \to \gamma_s^{\bar{u}}(x)$, where $(s,x) \to \gamma_s^{\bar{u}}(x)$ is the flow of some complete associated vectorfield $g_0 + \sum_{j=1}^m \bar{u}_j g_j$, $\bar{u} = (\bar{u}_1,\ldots,\bar{u}_m) \in \Omega$. The global definition of $W_v(t,\sigma,u)$ for each piecewise constant control u on $[0,\infty)$ can now be given as

(4.9) $\qquad W_v(t,\sigma,u)_{ij} = dH_i\big(\psi^u_{t,0}(x_0)\big)\big(\psi^u_{t,\sigma}\big)_* g_j\big(\psi^u_{\sigma,0}(x_0)\big) \qquad t,\sigma \ge 0$

where $(\psi^u_{t,\sigma})_* : T_{\psi^u_{\sigma,0}(x_0)}M \to T_{\psi^u_{t,0}(x_0)}M$ is the derivative of $\psi^u_{t,\sigma}$. Similarly

$W_a(t,\sigma,u)$ is globally defined as

(4.10)  $W_a(t,\sigma,u)_{ij} = -[(\psi^u_{\sigma,t})^* dH_j(\psi^u_{\sigma,0}(x_0))] g_i(\psi^u_{t,0}(x_0))$  $\quad$ $t,\sigma \geq 0$

where $(\psi^u_{\sigma,t})^*: T^*_{\psi^u_{\sigma,0}(x_0)}M \rightarrow T^*_{\psi^u_{t,0}(x_0)}M$ is the codifferential of $\psi^u_{\sigma,t}$. It follows that

(4.11)  $W_a(t,\sigma,u) = - W_v(\sigma,t,u)^T$  $\quad$ $t,\sigma \geq 0$

Remark If u is defined on $(-\infty,\infty)$ the same argument shows that $W_a(t,\sigma,u)$ and $W_v(t,\sigma,u)$ one defined for $t,\sigma \in (-\infty,\infty)$.

We now come to the definition of self-adjointness.

DEFINITION 4.1

A variational system (4.2) along a piecewise constant control u is called self-adjoint if the input-output map (4.5) of (4.2) for $v_0 = 0$ is equal to the input-output map (4.7) of the adjoint system (4.3) for $p_0 = 0$; i.e. if $u^a(\sigma) = u^v(\sigma)$ for $0 \leq \sigma \leq t$, then $y^a(\sigma) = y^v(\sigma)$, $0 \leq \sigma \leq t$, for any $t > 0$.

Remark Using Lemma 2.1 we see that self-adjointness is also equivalent to the following condition: for any variational control $u^v(t)=u^a(t)$ the time-evolutions $v(t)$, resp. $p(t)$, of the variational, resp. adjoint, system for $v_0 = p_0 = 0$ should satisfy $< p(t), v(t) > = 0$ for all $t \geq 0$.

From (4.5), (4.7) and (4.11) it follows that a variational system along u is self-adjoint if and only if $W_v(t,\sigma,u) = W_a(t,\sigma,u) = -W_v(\sigma,t,u)^T$, $t \geq \sigma \geq 0$, and hence if and only if

(4.12)  $W_v(t,\sigma,u) = -W_v(\sigma,t,u)^T,$  $\quad$ $t,\sigma \geq 0$

For later use we remark that $W_v(t,\sigma,u)$ may be decomposed into a product of a matrix only depending on t and a matrix depending on $\sigma$. As a matter of fact

$$W_v(t,\sigma,u)_{ij} = dH_i(\psi^u_{t,0}(x_0))(\psi^u_{t,0})^*(\psi^u_{0,\sigma})^* g_j(\psi^u_{\sigma,0}(x_0))$$

and so if we pick a local coordinate chart around $x_0$, and inparticular natural coordinates for $T_{x_0}M$ and $T^*_{x_0}M$ we may write

(4.13)  $W_v(t,\sigma,u) = G(t,u)H(\sigma,u)$

where $G(t,u)$ is the $m \times k$ matrix whose i-th row represents the covector in $T^*_{x_0}M$

(4.14) $\qquad (\psi_{t,0}^{u})^{*} dH_{i}(\psi_{t,0}^{u}(x_0))$

and $H(\sigma,u)$ is the $k \times m$ matrix whose $j$-th column represents the tangent vector in

$T_{x_0} M$

(4.15) $\qquad (\psi_{0,\sigma}^{u})_{*} g_{j}(\psi_{\sigma,0}^{u}(x_0))$

Then the self-adjointness condition can be equivalently stated as

(4.16) $\qquad G(t,u)H(\sigma,u) = -H^{T}(t,u)G^{T}(\sigma,u), \qquad t, \sigma \geq 0$

We now come to the main theorem of this chapter. As a preliminary remark we note that if for a certain u the variational system (4.2) is a _time-invariant_ system, i.e. $A(t)=A$, $B(t)=B$, $C(t)=C$ are all _constant_ matrices (this happens for instance if u=0 and $x_0$ is an equilibrium point), then self-adjointness amounts to the equality

(4.17) $\qquad Ce^{At}B = -B^{T}e^{-A^{T}t}C^{T}, \quad \forall t$

If we furthermore _assume_ that the linearized system $\dot{v} = Av + Bu^{v}$, $y^{v} = Cv$, is _minimal_, then it follows from Theorem 1.2 that this _linear_ system is actually Hamiltonian, i.e. $A^{T}J + JA = 0$, $B^{T}J = C$, for some non-singular antisymmetric matrix $J$(cf. [B2]). The following theorem shows how this extends to the nonlinear case.

THEOREM 4.2

Let (4.1) be a minimal nonlinear system. Then the system is Hamiltonian if and only if every variational system along any piecewise constant right continuous control u is self-adjoint.

Before proving Theorem 4.2 we note that the theorem can be equivalently stated in the following way. The input-output map $(u,u^{v}) \to (y,y^{v})$ of the _prolonged system_ (2.15) consists of the input-output map of the original system (4.1), together with the input-output map (4.5) of the variational system along u. Similarly, the input-output map $(u,u^{a}) \to (y,y^{a})$ of the _Hamiltonian extension_ (2.18) consists of the input-output map of the original system (4.1), together with the input-output map (4.7) of the adjoint system. Hence Theorem 4.2 can be rephrased as

THEOREM 4.2'

A minimal nonlinear system (4.1) is Hamiltonian if and only if the input-output map of its prolongation with initial state $x_p(0) = (x_0,0)$ coincides with the input-output map of its Hamiltonian extension with initial state $x_e(0) = (x_0,0)$.

First we prove the easy "only if" direction of Theorem 4.2.

Proof of Theorem 4.2 (==>) Consider a Hamiltonian system (not necessarily minimal) on $(M,\omega)$, together with its prolongation on $TM$ and Hamiltonian extension on $T^*M$. The symplectic form $\omega$ induces a natural bundle isomorphism $\bar{\omega}: TM \to T^*M$ defined by $\bar{\omega}(X) = \omega(x)(X,-)$, $X \in T_xM$. We will show that $\bar{\omega}$ is actually an isomorphism between the prolongation and the Hamiltonian extension, i.e. we will prove that $\bar{\omega}(x_p) = x_e$ along solutions of (2.15) and (2.18) respectively since

$$\bar{\omega}_* \dot{X}_0 = X_p^T X_0$$

(4.18)
$$\bar{\omega}_* \dot{X}_{H_j} = X_p^T X_{H_j} \qquad\qquad j = 1,\ldots,m$$

$$\bar{\omega}_*(-X_{H_j}^\ell) = X_{H_j^\ell}$$

$$p^T(-X_{H_j})\circ\bar{\omega} = \dot{H}_j$$

(Notice the minus signs in the last two equations. They appear as a result of the sign convention in the definition of a Hamiltonian system (2.2).) By Darboux's theorem there exist local coordinates $x = (q_1,\ldots,q_n,p_1,\ldots,p_n)$ for $M$ such that $\omega = \sum_{i=1}^{n} dp_i \wedge dq_i$. We may check (4.18) on every such a Darboux neighbourhood. In these coordinates (and the corresponding natural coordinates $(x,v)$ for $TM$ and $(x,p)$ for $T^*M$ where $p$ is not to be confused with $p_1,\ldots,p_n$ partial coordinates for $M$), the mapping $\bar{\omega}$ is given by

(4.19)
$$\bar{\omega}(x,v) = (x,Jv) = (x,p), \qquad \text{where } J = \begin{pmatrix} 0 & -I_n \\ I_n & 0 \end{pmatrix}$$

Therefore

$$
\bar{\omega}_* \dot{X}_0 = \begin{bmatrix} I_{2n} & 0 \\ 0 & J \end{bmatrix} \begin{pmatrix} X_0 \\ \frac{\partial X_0}{\partial x} v \end{pmatrix} \circ \bar{\omega}^{-1} = \begin{pmatrix} X_0 \\ J \frac{\partial X_0}{\partial x} v \end{pmatrix} \circ \bar{\omega}^{-1} =
$$

$$
\begin{bmatrix} X_0 \\ J \frac{\partial X_0}{\partial x} J^{-1} p \end{bmatrix} = \begin{pmatrix} X_0 \\ -(\frac{\partial X_0}{\partial x})^T p \end{pmatrix} = X_p^T X_0
$$

since $\left( \frac{\partial X_0}{\partial x} \right)^T J + J \left( \frac{\partial X_0}{\partial x} \right) = 0$ because $X_0$ is locally Hamiltonian.

Similarly $\bar{\omega}_* \dot{X}_{H_j} = X_p^T X_{H_j}$ . Furthermore

$$
\bar{\omega}_* X_{H_j}^\ell = \begin{pmatrix} I_{2n} & 0 \\ 0 & J \end{pmatrix} \begin{pmatrix} 0 \\ X_{H_j} \end{pmatrix} = \begin{pmatrix} 0 \\ JX_{H_j} \end{pmatrix} = - X_{H_j}^\ell
$$

and

$$
(p^T X_{H_j}) \circ \bar{\omega} = (Jv)^T X_{H_j} = -v^T JX_{H_j} = -(\frac{\partial H_j}{\partial x})v = - \dot{H}_j . \qquad \square
$$

For the proof of the "if" direction of Theorem 4.2 we need some intermediate steps.

<u>LEMMA 4.3</u>

Consider a minimal nonlinear system (4.1). Suppose the prolongation and the Hamiltonian extension have the same input-output map for the initial conditions $x_0$, $v_0 = p_0 = 0$. Then there exists a unique diffeomorphism (which is even real analytic) $\Phi : TM \rightarrow T^*M$ satisfying $\Phi(x_0,0) = (x_0,0)$ such that

$$
\Phi_* \dot{g}_i = X_p^T g_i \qquad i=0,1,\ldots,m
$$

$$
\Phi_* g_j^\ell = X_{H_j}^\ell \qquad j=1,\ldots,m
$$

(4.20)
$$
H_j^\ell \circ \Phi = H_j^\ell \qquad j=1,\ldots,m
$$

$$
H^{g_j} \circ \Phi = \dot{H}_j \qquad j=1,\ldots,m
$$

Moreover $\Phi$ is a bundle isomorphism which is the identity on the base manifold M, and so has in natural coordinates the form $\Phi(x,v)=(x,\phi(x,v))$ for a certain map $\phi$.

Proof: By Corollaries 3.3 and 3.7 the prolongation and the Hamiltonian extension are minimal. Furthermore it is easily seen that if g is a complete vectorfield on M, then $\dot{g}$ and $g^\ell$ are complete vectorfields on TM, and $X_p T_{g(x)}$ is a complete vectorfield on T*M. Moreover a vectorfield $X_\ell$ on T*M with H:M → R is trivially complete. Therefore we can invoke the Sussmann theorem ([S1, Theorem 5]) to conclude to the existence of a unique diffeomorphism $\phi$: TM → T*M satisfying $\phi(x_0,0) = (x_0,0)$ and (4.20). (Note that for the Sussmann theorem we actually only need accessibility instead of strong accessibility.)

Clearly the input-output behavior of the original system is not influenced by the behavior of the variational or adjoint system. By the minimality of the original system there exists a unique diffeomorphism from M into itself mapping $x_0$ onto $x_0$ and mapping the system into itself, namely the identity mapping. By uniqueness of $\phi$ it therefore follows that $\phi$ is of the form $\phi(x,v) = (x,\phi(x,v))$. □

LEMMA 4.4

Under the same assumptions as in Lemma 4.3, there exists a unique non-singular anti-symmetric matrix $\omega(x)$ depending on $x \in M$ such that $\phi(x,v) = \omega(x)v$, for all $v \in R^k$.

Proof It follows from $H^{g_j} \circ \phi = \dot{H}_j$ (see 4.20) and $\phi(x,v)=(x,\phi(x,v))$ that $g_j^T(x)\phi(x,v) = \frac{\partial H_j}{\partial x}(x)v$, $\forall v \in R^k$, $j=1,\ldots,m$. Furthermore from $\phi_* \dot{g}_i = X_p T_{g_i}(x)$ (see (4.20)) it follows that $(L_{X_p T_{g_i}(x)} H^{g_j}) \circ \phi = L_{\dot{g}_i} H_j$, or equivalently (see Lemma 3.2, 3.6) that $H^{[g_i,g_j]} \circ \phi = g_i(\dot{H}_j)$, $j=1,\ldots,m$, $i=0,1,\ldots,m$. Since $\phi(x,v) = (x,\phi(x,v))$ this yields

$$[g_i,g_j]^T(x)\phi(x,v) = \frac{\partial L_{g_i}(H_j)}{\partial x} v, \qquad v \in R^k.$$

In general for all $v \in R^k$

(4.21)    $[f_1,[f_2,[f_3\ldots[f_s,g_j]\ldots]]]^T(x)\phi(x,v) = \frac{\partial}{\partial x}(L_{f_1} L_{f_2} L_{f_3} \cdots L_{f_s} H_j)v$

with the $f_r$, $r = 1,\ldots,s$ equal to some $g_i$, $i = 0,1,\ldots,m$. Since the right hand side is linear in v and the system is minimal it follows that there exists a matrix $\omega(x)$ such that $\phi(x,v)= \omega(x)v$. Since $\phi$ is a diffeomorphism $\omega(x)$ is nonsingular for every x. It follows from (4.21) that $\omega(x)$ satisfies

(4.22) $\qquad [f_1,[f_2,[f_3\ldots[f_s,g_j]\ldots]]]^T(x)\omega(x) = \frac{\partial}{\partial x}(L_{f_1}L_{f_2}\ldots L_{f_s}H_j)$

with the $f_i$ as above. On the other hand in local coordinates $\Phi_*g_j^\ell = X_{H_j}^\ell$ together with $\Phi(x,v)=(x,\omega(x)v)$ yields

$$\begin{bmatrix} I & 0 \\ \frac{\partial}{\partial x}(\omega(x)v) & \omega(x) \end{bmatrix} \begin{bmatrix} 0 \\ g_j(x) \end{bmatrix} = \begin{bmatrix} 0 \\ -(\frac{\partial H_j}{\partial x})^T(x) \end{bmatrix}$$

or equivalently

(4.23) $\qquad \omega(x)\,g_j(x) = -(\frac{\partial H_j}{\partial x})^T(x) \qquad\qquad j=1,\ldots,m$

Furthermore by Lemma 3.2 and 3.6, and (4.20)

$$\Phi_*[g_i,g_j]^\ell = \Phi_*[\dot{g}_i,g_j^\ell] = [X_{p^Tg_i}, X_{H_j}^\ell] = X_{(L_{g_i}H_j)}^\ell$$

for $i=0,1,\ldots,m$, $j=1,\ldots,m$. As in (4.20) this yields in local coordinates

(4.24) $\qquad \omega(x)\,[g_i,g_j]\,(x) = -(\frac{\partial L_{g_i}H_j}{\partial x})^T(x)$

In general we obtain

(4.25) $\qquad \omega(x)\,[f_1,[f_2\ldots[f_s,g_j]\ldots]](x) = -(\frac{\partial}{\partial x}L_{f_1}L_{f_2}\ldots L_{f_s}H_j)^T(x)$

with the $f_r$, $r=1,\ldots,s$ equal to $g_i$, $i=0,1,\ldots,m$. Comparing (4.25) with (4.22), and invoking once more the minimality we conclude that $\omega(x)$ satisfies $\omega(x) = -\omega^T(x)$. Furthermore since $\omega(x)$ is non-singular and anti-symmetric it necessarily follows that $k = \dim M$ is even, say $k = 2n$. $\qquad\qquad\square$

## LEMMA 4.5

Let $\omega(x)$ be the $2n \times 2n$ matrix as in Lemma 4.4, under the same assumptions as in Lemma 4.3. Denote the $(i,j)$-th element of $\omega(x)$ by $\omega_{ij}(x)$. Then the two form $\omega := \sum_{i,j=1}^{2n} \omega_{ij}(x)\,dx_i \wedge dx_j$ is <u>closed</u> ($d\omega=0$), and so is a symplectic form on M.

<u>Proof</u> By minimality H contains locally $2n$ independent functions. Hence we can take

local coordinates $(x_1,\ldots,x_{2n})$ for M such that every coordinate function $x_i$ is of the form $L_{f_1}L_{f_2}\ldots L_{f_s}H_j$ for a certain $j \in \{1,\ldots,m\}$ and certain $f_r$, $r=1,\ldots,s$, equal to $g_i$, $i=0,1,\ldots,m$. It follows from (4.22) that there exist $2n$ independent vectorfields $k^1,\ldots,k^{2n}$ in $L_0$ of the form $[f_1,[f_2\ldots[f_s,g_j]\ldots]]$ such that

(4.26) $\qquad (k^i)^T(x)\omega(x) = \dfrac{\partial x_i}{\partial x} = e_i^{\ T} \qquad i=1,\ldots,2n$

with $e_i$ the i-th basis vector of $R^{2n}$. Denote the component functions of $k^i$ by $k^{i1},\ldots,k^{i2n}$. Now define the $2n \times 2n$ matrix $K(x)$ with $(i,j)$-th element equal to $k^{ij}(x)$. It follows from (4.26) that

(4.27) $\qquad K(x)\omega(x) = I_{2n}$

Furthermore we know from (4.20) that

(4.28) $\qquad \phi_* \dot{k}^i = X_{p^T k^i} \qquad\qquad i=1,\ldots,2n$

In local coordinates $(x_1,\ldots,x_{2n})$ this yields

$$\begin{bmatrix} I & 0 \\ \dfrac{\partial}{\partial x}(\omega(x)v) & \omega(x) \end{bmatrix} \begin{bmatrix} k^i(x) \\ \dfrac{\partial k^i}{\partial x}(x)v \end{bmatrix} = \begin{bmatrix} k^i(x) \\ -\left(\dfrac{\partial k^i}{\partial x}\right)^T(x)p \end{bmatrix} \circ \phi(x,v)$$

or equivalently

(4.29) $\qquad \dfrac{\partial}{\partial x}(\omega(x)v)k^i(x) + \omega(x)\dfrac{\partial k^i}{\partial x}(x)v + \left(\dfrac{\partial k^i}{\partial x}\right)^T(x)\omega(x)v = 0 \qquad$ for all $v$

Writing out for $v = (v_1,\ldots,v_{2n})^T$

$$\dfrac{\partial}{\partial x}(\omega(x)v) = \left( \sum_{s=1}^{2n} \dfrac{\partial \omega_{ks}}{\partial x_\ell} v_s \right) \qquad k,\ \ell = 1,\ldots,2n$$

and so the k-th component of $\dfrac{\partial}{\partial x}(\omega(x)v)k^i(x)$ is equal to

$$\sum_{\ell=1}^{2n} \left( \sum_{s=1}^{2n} \dfrac{\partial \omega_{ks}}{\partial x_\ell} v_s \right) k^{i\ell} = \sum_{s=1}^{2n} \left( \sum_{\ell=1}^{2n} \dfrac{\partial \omega_{ks}}{\partial x_\ell} k^{i\ell} \right) v_s$$

Furthermore the $\ell$-th component of $\dfrac{\partial k^i}{\partial x}(x)v$ is given by $\displaystyle\sum_{s=1}^{2n} \dfrac{\partial k^{i\ell}}{\partial x_s} v_s$, and so the k-th component of $\omega(x)\dfrac{\partial k^i}{\partial x}(x)v$ by

$$\sum_{\ell=1}^{2n} \omega_{k\ell} \left( \sum_{s=1}^{2n} \frac{\partial k^{i\ell}}{\partial x_s} v_s \right) = \sum_{s=1}^{2n} \left( \sum_{\ell=1}^{2n} \omega_{k\ell} \frac{\partial k^{i\ell}}{\partial x_s} \right) v_s \, .$$

Finally the k-th component of $\left(\frac{\partial k^1}{\partial x}\right)^T (x) \omega(x) v$ is given by

$$\sum_{\ell=1}^{2n} \frac{\partial k^{i\ell}}{\partial x_k} \left( \sum_{s=1}^{2n} \omega_{\ell s} v_s \right) = \sum_{s=1}^{2n} \left( \sum_{\ell=1}^{2n} \omega_{\ell s} \frac{\partial k^{i\ell}}{\partial x_k} \right) v_s$$

Since $v = (v_1, .., v_{2n})^T$ is arbitrary we therefore obtain from (4.29)

(4.30) $$\sum_{\ell=1}^{2n} \frac{\partial \omega_{ks}}{\partial x_\ell} k^{i\ell} + \sum_{\ell=1}^{2n} \omega_{k\ell} \frac{\partial k^{i\ell}}{\partial x_s} + \sum_{\ell=1}^{2n} \frac{\partial k^{i\ell}}{\partial x_k} \omega_{\ell s} = 0$$

Furthermore by differentiation of (4.27) we have

(4.31) $$\sum_{\ell=1}^{2n} \omega_{i\ell} \frac{\partial k^{\ell j}}{\partial x_s} + \sum_{\ell=1}^{2n} \frac{\partial \omega_{i\ell}}{\partial x_s} k^{\ell j} = 0 \qquad i, j, s = 1, .., 2n$$

Hence $$\sum_{\ell=1}^{2n} \omega_{k\ell} \frac{\partial k^{i\ell}}{\partial x_s} = - \sum_{\ell=1}^{2n} \frac{\partial \omega_{k\ell}}{\partial x_s} k^{i\ell} = \sum_{\ell=1}^{2n} \frac{\partial \omega_{\ell k}}{\partial x_s} k^{i\ell},$$

and $$\sum_{\ell=1}^{2n} \omega_{\ell s} \frac{\partial k^{i\ell}}{\partial x_k} = - \sum_{\ell=1}^{2n} \omega_{s\ell} \frac{\partial k^{i\ell}}{\partial x_k} = \sum_{\ell=1}^{2n} \frac{\partial \omega_{s\ell}}{\partial x_k} k^{i\ell}.$$

Insertion in (4.30) yields

(4.32) $$\sum_{\ell=1}^{2n} \left( \frac{\partial \omega_{ks}}{\partial x_\ell} k^{i\ell} + \frac{\partial \omega_{\ell k}}{\partial x_s} k^{i\ell} + \frac{\partial \omega_{s\ell}}{\partial x_k} k^{i\ell} \right) = 0$$

By non-singularity of the matrix K(x) we therefore obtain

(4.33) $$\frac{\partial \omega_{ks}}{\partial x_\ell} + \frac{\partial \omega_{\ell k}}{\partial x_s} + \frac{\partial \omega_{s\ell}}{\partial x_k} = 0 \qquad \ell, k, s = 1, ..., 2n$$

Now equations (4.33) form exactly the local coordinate expressions for the closedness of the two-form

$$\omega := \sum_{i,j=1}^{2n} \omega_{ij} dx_i \wedge dx_j \text{ on } M$$

(see for instance [A, Chapter 2]).  □

Finally we prove the "if" direction of Theorem 4.2.

Proof of Theorem 4.2 (<==) In Lemma 4.3, 4.4, 4.5 we have deduced the existence of a symplectic form $\omega := \sum_{i,j=1}^{2n} \omega_{ij}(x) dx_i \wedge dx_j$ on M. Furthermore from (4.20) or (4.23) it follows that for $j=1,\ldots,m$, $g_j^T(x)\omega(x) = \dfrac{\partial H_j}{\partial x}(x)$. Hence $g_j = -X_{H_j}$, $j=1,\ldots,m$. It remains to be proved that $g_0$ is a locally Hamiltonian vectorfield. Now by Darboux's theorem we can take local coordinates $x = (q_1,\ldots,q_n,p_1\ldots,p_n)$ for M such that locally $\omega = \sum_{i=1}^{n} dp_i \wedge dq_i$. In such coordinates $\Phi_* \dot{g}_0 = X_p^T g_0$ amounts to

$$\begin{bmatrix} I & 0 \\ 0 & J \end{bmatrix} \begin{bmatrix} g_0 \\ \dfrac{\partial g_0}{\partial x}(x)v \end{bmatrix} = \begin{bmatrix} g_0 \\ -(\dfrac{\partial g_0}{\partial x})^T(x)Jv \end{bmatrix}$$

or equivalently

(4.34)  $$J \frac{\partial g_0}{\partial x}(x) + \left(\frac{\partial g_0}{\partial x}\right)^T(x)J = 0$$

where of course $J = \begin{pmatrix} 0 & -I_n \\ I_n & 0 \end{pmatrix}$. Hence the Jacobian matrix $\dfrac{\partial}{\partial x}(g_0^T J)(x) = (\dfrac{\partial g_0}{\partial x})^T(x)J$ is symmetric, so there exists on this Darboux neighbourhood a function $H_0$ for which

(4.35)  $$g_0^T(x)J = -\frac{\partial H_0}{\partial x}(x).$$

Hence on this neighbourhood $g_0 = X_{H_0}$.  □

Remark Note that in general it is hard to give conditions in order that the system is globally Hamiltonian, i.e. in order that the internal energy $H_0$ is globally defined. (Except for the case that M is such that every closed one-form is exact, e.g. if M is simply connected. However since we insist on observability simple-connectedness is not a natural assumption, cf.[C4].)

Let us briefly return to Lemma 2.1. It follows from the proof of Theorem 4.2 ("if"-direction) that if $p(t)$ is a solution of the adjoint system for an input $u^a(t)$, then $p(t)$ equals $\omega(x(t))v(t)$, where $v(t)$ is the solution of the variational system with input $u^v(t) = u^a(t)$ and initial condition $v_0 = (\omega(x_0))^{-1}p_0$. Hence for a Hamiltonian system, (2.6) becomes

$$(4.36) \qquad \frac{d}{dt}\omega_{x(t)}(v_1(t),v_2(t)) = (u_2^v(t))^T y_1^v(t) - (u_1^v(t))^T y_2^v(t)$$

where $v_i(t)$ is the solution of the variational system for the input $u_1^v(t)$ and with output $y_i^v(t)$, $i=1,2$. (Recall that for a Hamiltonian system the input-output maps of the variational and adjoint system coincide.) This formula was already proved in [V1] by direct methods, and will serve as a starting point for the characterization of Hamiltonian systems completely in terms of their variational input-output behavior in the next chapter.

Finally, in previous work [V1,G1,V3,V5] it has been shown that minimal Hamiltonian systems with the same input-output map are not only diffeomorphic, as implied by the Sussmann uniqueness theorem [S1], but also symplectomorphic. This fact also follows immediately from the proof of Theorem 4.2, as we will now briefly show. Let $\Sigma_1$ and $\Sigma_2$ be two minimal Hamiltonian systems with state spaces $(M_1,\omega_1)$, respectively $(M_2,\omega_2)$. Then, as in Theorem 4.2, there exist maps $\Phi_i: TM_1 \to T^*M_1$, $i=1,2$, of the form $\Phi_i(x,v) = (x,\omega_1(x)v)$, $i=1,2$, where $\omega_i(x)$, $i=1,2$, are the matrices corresponding to the symplectic forms $\omega_i$, $i=1,2$. Now let $\Sigma_1$ and $\Sigma_2$ have the same input-output map for groundstates $x_0^1 \in M^1$ and $x_0^2 \in M^2$. By the Sussmann uniqueness theorem there exists a diffeomorphism $\psi: M_1 \to M_2$ transforming $\Sigma_1$ into $\Sigma_2$. Clearly, also the prolongations of $\Sigma_1$ and $\Sigma_2$ have the same input-output maps, as well as the Hamiltonian extensions of $\Sigma_1$ and $\Sigma_2$. It is easily seen that the equivalence mapping between the two prolongations is given by $\psi_*: TM_1 \to TM_2$, and between the two Hamiltonian extensions by $(\psi^*)^{-1}: T^*M_1 \to T^*M_2$. By the uniqueness of these equivalence mappings we therefore obtain the commutative diagram

$$(4.37)$$

$$
\begin{array}{ccc}
TM_1 & \xrightarrow{\psi_*} & TM_2 \\
\Phi_1 \downarrow & & \downarrow \Phi_2 \\
T^*M_1 & \xleftarrow{\psi^*} & T^*M_2
\end{array}
$$

i.e.,

$$(4.38) \qquad \psi^* \circ \Phi_2 \circ \psi_* = \Phi_1.$$

Since $\Phi_i(x,v) = (x,\omega_i(x)v)$, $i=1,2$, we see that (4.38) is equivalent to

(4.39)     $\psi^{*}\omega_2 = \omega_1.$

so $\psi$ is a symplectomorphism. Note furthermore that it follows from (4.39) that the internal energies $H_0^1$, i=1,2, of $\Sigma_1$ and $\Sigma_2$ on their domains of definition are related by (cf.[V1])

(4.40)     $H_0^2 \circ \psi = H_0^1 + \text{constant}.$

i.e., the internal energy of a minimal Hamiltonian system is, up to a constant, uniquely determined by its input-output map.

# 5. THE VARIATIONAL CRITERION

The main purpose of this chapter is to establish the exact relationship between
self adjointness of the variational systems corresponding to a minimal system, and
the criterion conjectured by Van der Schaft, as described in the chapter (1). We
also collect together the main results of the monograph and present them in a
unified way. Finally we demonstrate formally that the external behaviour set of a
minimal Hamiltonian system is a Lagrangian submanifold of the manifold of all
behaviours from minimal systems.

<u>SECTION 5.1</u>

In this chapter we deal exclusively with complete, affine and analytic systems

(5.1)
$$\dot{x} = g_0(x) + \sum_{i=1}^{m} u_i g_i(x), \quad x \in M, \quad x(0) = x_0$$

$$y_i = H_i(x), \quad 1 \leq i \leq m, \quad u \in \Omega \subset R^m.$$

We shall use the definitions in chapter (1.6) of $\Sigma_i^+(0)(x_0)$, $\Sigma_e^+(0)(x_0)$,
variations of elements in these behaviour sets, and the corresponding variational
fields. If $(\bar{u}, \bar{y}, \bar{x}) \in \Sigma_i^+(0)(x_0)$, and $t \longrightarrow \delta u(t)$ is any piecewise constant right
continuous $R^m$ valued function defined on $[0,\infty)$ with support contained in $(0,\infty)$ we
define a variation of $\bar{u}$ by setting

$$u(t,\varepsilon) = \bar{u}(t) + \varepsilon \, \delta u(t).$$

It follows that $t \dashrightarrow u(t,\varepsilon)$ is piecewise constant for each $\varepsilon$, but it defines  an
admissible input to (5.1) only if

(5.2)
$$\bar{u}(t) + \varepsilon \, \delta u(t) \in \Omega, \quad t \in (0,\infty)$$

Supposing that this condition is satisfied for each $\varepsilon$ in a sufficiently small
neighbourhood $V$ of $0$, we obtain corresponding variations $y(t,\varepsilon)$ of $\bar{y}(t)$ and
$x(t,\varepsilon)$ of $\bar{x}(t)$. That for each $\varepsilon \in V$ we have $t \rightarrow (u(t,\varepsilon), y(t,\varepsilon), x(t,\varepsilon)) \in$
$\Sigma_i^+(0)(x_0)$, follows from the completeness assumption of (5.1), equation (5.2) and
the fact that the support of $\delta u$ is contained in $(0,\infty)$. It follows from the smooth
dependence of solutions of differential equations on parameters that the
variational field along $(\bar{u}, \bar{y}, \bar{x})$, $t \longrightarrow (\delta u(t), \delta y(t), \delta x(t))$ exists and satisfies
the required regularity conditions. It follows that we may realize any piecewise
constant Rm valued function with support contained in $(0,\infty)$ as the control

component of the variational field along any $(\bar{u}, \bar{y}, \bar{x}) \in \Sigma_i^+ (0)(x_0)$, as long as (5.2) is satisfied for $|\varepsilon|$ sufficiently small. We shall always assume that this is the case, which for example occurs when $\Omega = \mathbb{R}^m$. We now give the definition of admissible variations. We shall write supp f, for the support of f, which is the smallest closed set outside which f vanishes. Also recall that if $(\delta u, \delta y)$ is a variational field along $(u,y) \in \Sigma_e^+(0)(x_0)$ then $\delta u(0) = 0$.

## DEFINITION 5.1

An admissible variation $(u,y)$ of $(\bar{u}, \bar{y}) \in \Sigma_e^+ (0)(x_0)$ is a variation which satisfies
(i) $\delta u$ is piecewise constant and supp $\delta u$ is compact. Thus there exists $T_1$, $T_2 > 0$ such that supp $(u(\cdot, \varepsilon) - \bar{u}(\cdot)) \subset [T_1, T_2]$ for every $\varepsilon$ sufficiently small.
(ii) supp $\delta y \subset$ supp $\delta u$.
(iii) Suppose supp $\delta u \subset (0, T)$, and $(\bar{u}', \bar{y}') \in \Sigma_e^+ (0)(x_0)$ are such that $\bar{u}'(t) = \bar{u}(t)$ (and hence $\bar{y}'(t) = \bar{y}(t)$) for $t \in [0, T]$). Define a map $(t, \varepsilon) \rightarrow u'(t, \varepsilon)$ by setting $u'(t, \varepsilon) = u(t, \varepsilon)$, $t \in [0, T]$, $u'(t, \varepsilon) = \bar{u}'(t)$, $t \in (T, \infty)$. We deduce that u' is a variation of $\bar{u}'$, and hence obtain a corresponding variation $(u', y')$ of $(\bar{u}', \bar{y}')$. We require the corresponding variational field $(\delta u', \delta y')$ along $(\bar{u}', \bar{y}')$ to satisfy condition (ii) also. Clearly we have $\delta u = \delta u'$.

From now on we shall abuse notation and simply say $(\delta u, \delta y)$ is an (admissible) variation of $(\bar{u}, \bar{y}) \in \Sigma_e^+ (0)(x_0)$ (with compact support), when $(\delta u, \delta y)$ is the variational field of an (admissible) variation $(u,y)$ of $(\bar{u}, \bar{y})$. Note that part of the condition (ii) may be written as follows: - if $\delta u = 0$ on $(0, T_1)$ then $\delta y = 0$ on $(0, T_1)$ also. For variations $(\delta u, \delta y)$ of $(\bar{u}, \bar{y}) \in \Sigma_e^+(0)(x_0)$ this condition is automatically satisfied by causality. On the other hand the other part of condition (ii), expressed as: - if $\delta u = 0$ on $[T_2, \infty)$ then $\delta y = 0$ on $[T_2, \infty)$ also, is a definite constraint. Loosely stated definition (5.1) may be expressed as:
- $(\delta u, \delta y)$ is an admissible variation of $(\bar{u}, \bar{y}) \in \Sigma_e^+ (0)(x_0)$, if $(\delta u, \delta y)$ has compact support and this is independent of the values of $(\bar{u}, \bar{y})$ occuring to the right of the support of $(\delta u, \delta y)$.
Before proceeding with a discussion of the existence of admissible variations we need a technical result.

## LEMMA 5.2

If D is a dense subspace of a Hilbert space H, and S is a finite dimensional subspace of H, then $D \cap S^\perp$ is a dense subspace of $S^\perp$, with respect to its natural Hilbert space structure induced from H, where $S^\perp$ is the orthogonal complement of S.

Proof If $0 \neq a \in H$, we first prove that $D \cap a^\perp$ is dense in $a^\perp$. Note that $a^\perp$ is a codimension one closed subspace of $H$. Let $s \in a^\perp$ and let $\{d_n\} \subset D$ be a sequence such that $d_n \longrightarrow s$ as $n \longrightarrow \infty$. We may write $d_n = \alpha_n a + s_n$ where $\alpha_n \in R$ and $s_n \in a^\perp$, with $\alpha_n \to 0$ as $n \to \infty$. Now there exists $\beta \neq 0$, $\beta \in R$, such that $(\beta a + a^\perp) \cap D \neq \emptyset$, since otherwise $D \subset a^\perp$, which contradicts the fact that $D$ is dense in $H$. Thus there exists $r \in a^\perp$ and $\beta \neq 0$ such that $\beta a + r = d \in D$. Now

$$\bar{d}_n = d_n - d\,(\alpha_n/\beta) = s_n - r\,(\alpha_n/\beta) \in D \cap a^\perp$$

since $D$ is a subspace. However since $\alpha_n \to 0$ we have $\bar{d}_n \to s$ as $n \to \infty$, which shows that $D \cap a^\perp$ is dense in $a^\perp$. Now since $S$ is a finite dimensional subspace we may write $S = \text{span}\,\{a_1 \cdots a_n\}$ where $a_1, \cdots, a_n$ is an orthonormal basis of $S$. From the result above it follows that $D \cap a_1^\perp$ is dense in $a_1^\perp$. But $a_1^\perp = \text{span}\,\{a_2, \ldots, a_n\} + S^\perp$. Thus $(D \cap a_1^\perp) \cap a_2^\perp$ is dense in $a_2^\perp$ viewed as subspaces of $a_1^\perp$. We deduce that $D \cap (a_1^\perp \cap a_2^\perp)$ is dense in $\text{span}\,\{a_3, \cdots, a_n\} + S^\perp$. By a simple induction argument we obtain for any $r$, $1 \le r \le n - 1$, $D \cap (a_1^\perp \cap \cdots \cap a_r^\perp)$ is dense in $\text{span}\,\{a_{r+1}, \cdots, a_n\} + S^\perp$ and for $r = n$, $D \cap (a_1^\perp \cap \cdots \cap a_n^\perp)$ is dense in $S^\perp$. But $a_1^\perp \cap \cdots \cap a_n^\perp = S^\perp$, so $D \cap S^\perp$ is dense in $S^\perp$ as required.    []

Referring to the notation and definitions of chapters (2) and (3) we recall that a system (5.1) is called quasi-minimal if it is strongly accessible and weakly observable. We now have the following existence result for admissible variations.

## PROPOSITION 5.3

Consider a quasi-minimal, analytic and complete system (5.1). Given any $(u,y) \in \Sigma_e^+(0)(x_0)$, and $T > 0$ there exists a dense set of piecewise constant functions $\delta u$, in $L_2([0,T]; R^m) \cap S^\perp$, for some finite dimensional subspace $S \subset L_2([0,T]; R^m)$, which may be realized as a component of an admissible variation $(\delta u, \delta y)$ of $(u,y)$, with compact support contained in $(0,T)$. Furthermore if $(u,y,x)$ is the corresponding element of $\Sigma_1^+(0)(x_0)$, the corresponding variation $(\delta u, \delta y, \delta x)$ of $(u,y,x)$ is such that $\text{supp } \delta x \subset (0,T)$ also.

Proof Referring to chapters 2 and 4 we may write the variation $\delta x$ due to the variation $\delta u$ in the following manner, where $(t, \sigma, x) \to \psi_{t,\sigma}^u(x)$ is the flow of the (time-varying) vector field $g_0(t) + \sum_{i=1}^m u_i(t)\, g_i(x)$,

$$(5.3) \qquad \delta x(t) = \sum_{i=1}^m (\psi_{t,0}^u)_* \left( \int_0^t (\psi_{0,\sigma}^u)_* \, g_i(\psi_{\sigma,0}^u(x_0)) \, \delta u_i(\sigma) d\sigma \right).$$

We claim that if, as in the proposition, $(\delta u, \delta y)$ is an admissible variation of $(u, y)$ with compact support contained in $(0, T)$ then

(5.4) $$0 = \sum_{i=1}^{m} \int_0^T (\psi_{0,\sigma}^u)_* g_i(\psi_{\sigma,0}^u(x_0)) \delta u_i(\sigma) d\sigma.$$

From (5.3) and (5.4) we deduce that if this is the case then $\delta x(T) = 0$, and moreover since outside the interval $[0, T]$, $\delta u(t) = 0$, the structure of the variational equations (2.5) guarantees that $\delta x(t) = 0$ for $t \geq T$. It follows, modulo our claim, that the final statement of the proposition is indeed valid. Furthermore, if $\delta u$ is a variation of $u$ with compact support contained in $(0, T)$, which satisfies (5.4), it follows from the equations

$$\delta y_i(t) = dH_i(\psi_{t,0}^u(x_0)) \, \delta x(t), \quad 1 \leq i \leq m,$$

that the corresponding variation $\delta y$ of $y$ has compact support contained in $(0, T)$. It follows that the pair $(\delta u, \delta y)$ meets conditions (i) and (ii) of definition (5.1). Moreover if $(u', y') \in \Sigma_e^+(0)(x_0)$ satisfies $u'(t) = u(t)$, $y'(t) = y(t)$ for $t \in [0, T]$, then equations (5.3) and (5.4) are unchanged for $t \in [0, T]$. Thus if $\delta u' = \delta u$ and $(\delta u', \delta y', \delta x')$ is the corresponding variation of $(u', y', x')$, then $\delta x'(t) = 0$ for $t \in [T, \infty)$ and so $\delta y'$ has compact support contained in $(0, T)$ also. Thus condition (iii) of definition (5.1) is also met. We have therefore shown, modulo our claim, that the existence of admissible variations $(\delta u, \delta y)$ of $(u, y)$ with compact support contained in $(0, T)$ depends only on finding piecewise constant functions $\delta u$ on $[0, T]$, with compact support in $(0, T)$, which satisfy equation (5.4).

Fixing a local coordinate chart for $M$ about $x_0$ we may write equations (5.4) in the form

(5.5) $$0 = \int_0^T H(\sigma, u) \, \delta u(\sigma) d\sigma.$$

where $H(\sigma, u)_{ij}$ is the i'th component of the vector $(\psi_{0,\sigma}^u)_* g_j(\psi_{\sigma,0}^u(x_0)) \in T_{x_0} M$ in local coordinates. This may be expressed in terms of the Hilbert space $L_2([0, T]; \mathbb{R}^m)$ as the orthogonality of $\delta u$ with the subspace $S$ defined by

(5.6) $$S = \{H(\cdot, u)^T \alpha \; ; \; \alpha \in \mathbb{R}^n\}.$$

If $D$ is the dense subspace of $L_2([0, T]; \mathbb{R}^m)$ consisting of piecewise constant functions on $[0, T]$, with compact support contained in $(0, T)$, we see that admissible variations $(\delta u, \delta y)$ of $(u, y)$ are in one to one correspondence with the set of functions in $D \cap S^{\perp}$. However by lemma (5.2) $D \cap S^{\perp}$ is dense in $S^{\perp}$, and inparticular non empty.

It remains to verify our claim. Assume to the contrary that

(5.7) $$0 \neq v = \sum_{i=1}^{m} \int_{0}^{t} (\psi_{0,\sigma}^{u})_* \ g_i(\psi_{\sigma,0}^{u}(x_0)) \ \delta u_i(\sigma) d\sigma.$$

From proposition (3.8), and the quasi-minimality of (5.1) it follows that the following truncated prolongation is weakly observable.

$$\dot{x}_p = \dot{g}_0(x_p) + \sum_{i=1}^{m} u_i \ \dot{g}_i(x_p), \qquad x_p \in TM$$

(5.8) $$y_j = H_j^{\ell}(x_p), \qquad 1 \le j \le m$$

$$y_j^{v} = \dot{H}_j(x_p), \qquad 1 \le j \le m.$$

Inparticular there exists a piecewise constant control $\bar{u}$ on $[0,\infty)$, for which $y^v(t)$ is not identically zero on $[0,\infty)$, when the system is initialized at time zero at the state $\bar{x}_p = (\psi_{T,0}^{u}(x_0), (\psi_{T,0}^{u})_* v) \in TM$. Otherwise the initial states $\bar{x}_p$ and $(\psi_{T,0}^{u}(x_0), 0)$ for system (5.8) would be indistinguishable, which as in section 3 would contradict the fact that system (5.8) is weakly observable.

Now define a control $u'$ on $[0,\infty)$ for system (5.1) by setting $u'(t) = u(t)$, $t \in [0,T), u'(t) = \bar{u}(t-T)$ for $t \ge T$, and hence obtain a pair $(u', y') \in \Sigma_e^+(0)(x_0)$ as in (iii) of Definition (5.1). Since $(\delta u, \delta y)$ is admissible we may apply the conclusion of (iii) to see that the resulting variation $(\delta u', \delta y')$ also has compact support in $(0,T)$, where $\delta u' = \delta u$. Applying $\delta u' = \delta u$ to the variational system of (5.1) along $u'$, at time $T$ we reach the variational state (see equations (5.3)) $(\psi_{T,0}^{u'})_* v \in T_{\psi_{T,0}^{u'}(x_0)} M$. However for $t > T$ the output $\delta y(t)$ of the variational equation along $u'$, now coincides with the output $y^v(t)$ of system (5.8), initialized at $\bar{x}_p$. By construction this is not identically zero, contradicting the fact that supp $\delta y' \subset (0,T)$. We conclude that $v = 0$, establishing our claim. $\square$

SECTION 5.2

We now prove some intermediate results from which we deduce our main results.

LEMMA 5.4

Consider a quasi-minimal, analytic and complete system which is (locally) Hamiltonian; i.e. given by equations (1.15). Given any $(u,y) \in \Sigma_e^+(0)(x_0)$, and admissible variations $(\delta_i u, \delta_i y)$ of $(u,y)$ with compact support, $i = 1,2$, we have

$$\int_{0}^{\infty} (\delta_2 y(t)^T \delta_1(t) - \delta_1 y(t)^T \delta_2 u(t)) \ dt = 0.$$

Proof Suppose that the support of $(\delta u, \delta y)$ is contained in $(0, T)$. By proposition (5.3) if $(\delta u, \delta y, \delta x)$ is the corresponding variation of $(u, y, x) \in \Sigma_1^+ (0)(x_0)$, then supp $\delta x \subset (0, T)$. By equation (4.36), or Van der Schaft [V1],

$$\frac{d}{dt} \omega(x(t)) (\delta_1 x(t), \delta_2 x(t)) = \delta_2 y(t)^T \delta_1 u(t) - \delta_1 y(t)^T \delta_2 u(t)$$

where $\omega$ is the symplectic form associated with the Hamiltonian system (1.15). Thus

$$\int_0^\infty (\delta_2 y(t)^T \delta_1 u(t) - \delta_1 y(t)^T \delta_2 u(t)) \, dt$$

$$= \omega(x(T))(\delta_1 x(T), \delta_2 x(T)) - \omega(x_0)(\delta_1 x(0), \delta_2 x(0)) = 0.$$

[]

We recall from chapter (4) the definition of the kernel function $W_v(t, \sigma, u)$ which defines the response of the variational system along u. Moreover assuming, as we do always, that the system (5.1) has a state space M of dimension k, we may select a coordinate chart about $x_0$ and factor $W_v(t, \sigma, u)$ as $G(t,u)H(\sigma,u)$, where $G(\cdot, u)$ is an m × k matrix valued function for each control u. In terms of this factorization we may define

(5.9)
$$K^A(t, \sigma, u) = \begin{cases} G(t,u) \, H(\sigma,u) + H(t,u)^T G(\sigma,u)^T, & t > \sigma \\ -G(t,u) \, H(\sigma,u) - H(t,u)^T G(\sigma,u)^T, & t < \sigma \end{cases}$$

## PROPOSITION 5.5

Consider a quasi-minimal, analytic and complete system (5.1). Suppose that for any $(u,y) \in \Sigma_e^+ (0)(x_0)$, all admissible variations $(\delta_i u, \delta_i y)$ of $(u,y)$ with compact support, $i = 1, 2$, satisfy

(5.10)
$$\int_0^\infty (\delta_2 y(t)^T \delta_1 u(t) - \delta_1 y(t)^T \delta_2 u(t)) \, dt = 0.$$

Then there exists a matrix valued function $\tilde{G}$ such that for $t, \sigma \geq 0$

(5.11)
$$\tilde{G}(t,u)H(\sigma,u) - H(t,u)^T \tilde{G}(\sigma,u)^T = K^A(t, \sigma, u).$$

## Proof

We fix $T > 0$ and show that (5.11) is true for any $t, \sigma \in [0, T]$. As in proposition (5.3) the constraints on admissible variations $(\delta_1 u, \delta_1 y)$ of $(u,y)$ with support contained in $(0, T)$ may be expressed by equation $(5.5)$ or

$$0 = \int_0^T H(t,u) \, \delta_i u(t) dt , \qquad i = 1, 2.$$

By substituting the relationships between $\delta_i y$ and $\delta_i u$, namely

$$\delta_i y(t) = \int_0^t W_v(t,\sigma,u) \delta_i u(\sigma) d\sigma \qquad i = 1, 2,$$

into expression $(5.10)$ we obtain

$(5.12)$
$$0 = \int_0^T \int_0^t (\delta_1 u(t)^T W_v(t,\sigma,u) \delta_2 u(\sigma) - \delta_2 u(t)^T W_v(t,\sigma,u) \delta_1 u(\sigma)) d\sigma dt$$

However the constraints above may be expressed as

$$\int_0^t H(t,u) \delta_i u(t) dt = -\int_t^T H(t,u) \delta_i u(t) dt,$$

which when substituted into $((5.12)$, remembering that $W_v(t,\sigma,u) = G(t,u) H(\sigma,u)$, yields after some manipulation the following identities

$$0 = \int_0^T \int_0^t \delta_1 u(t)^T (W_v(t,\sigma,u) + W_v(\sigma,t,u)^T) \, \delta_2(\sigma)^T d\sigma dt$$

$$- \int_0^T \int_t^T \delta_1 u(t)^T (W_v(t,\sigma,u) + W_v(\sigma,t,u)^T) \, \delta_2 u(\sigma)^T d\sigma dt$$

These identities in turn yield the following expression

$(5.13)$
$$0 = \int_0^T \int_0^T \delta_1 u(t)^T K^A(t,\sigma,u) \, \delta_2 u(\sigma) d\sigma dt$$

To prove the identity $(5.11)$ from this, we give a Hilbert space setting to our situation. Let $H$ be the Hilbert space $L_2 ([0,T];R^m)$ consisting of $R^m$ valued functions on $[0,T]$, with inner product $\langle f,g \rangle = \int_0^T f(t)^T g(t) \, dt$, and let $S$ be the subspace $\{H(\cdot,u)^T \alpha; \, \alpha \in R^n\}$. Let $H_1$ be the Hilbert space $L_2([0,T] \times [0,T];R^{m^2})$ consisting of $m \times m$ matrix valued functions on $[0,T] \times [0,T]$ with inner product

$$\langle f,g \rangle = \int_0^T \int_0^T \text{trace} \, (f(t,\sigma)g(t,\sigma)^T) d\sigma dt.$$

Note $H_1$ may be viewed as the closure (in $H_1$) of the subspace consisting of all finite linear combinations of elements in $H \otimes H$. (If $f,g \in H$ then $f \otimes g \in H \otimes H$ is the function in $H_1$ given by $(t,\sigma) \rightarrow f(t) g(t)^T$.) Let $D$ be the dense subspace of

H, consisting of all piecewise constant functions in H with support contained in $(0, T)$. Similarly we let $D_1$ be the dense subspace of $H_1$ consisting of all finite linear combinations of elements in D ⊗ D.

Let $S_1$ be the subspace of $H_1$, consisting of all finite linear combinations of elements in S ⊗ H and H ⊗ S. $S_1$ may be identified with the space of all matrix valued functions of the form

$$H(t,u)^T K_1(\sigma)^T + K_2(\sigma) H(\sigma,u)$$

where $K_1$ and $K_2$ are $m \times k$ matrix valued functions with components in $L_2([0,T])$. Now the orthogonal complement of $S_1$ in $H_1$, denoted $S_1^\perp$, is just the closure in $H_1$ of the subspace consisting of all finite linear combinations of elements in $S^\perp \otimes S^\perp$. Thus $D_1 \cap S_1^\perp$ is the subspace consisting of all finite linear combinations of elements in $D \cap S^\perp \otimes D \cap S^\perp$. But by lemma (5.2) $D \cap S^\perp$ is dense in $S^\perp$, so $D_1 \cap S_1^\perp$ is dense in $S_1^\perp$. Noting that if A is an $m \times m$ matrix, and a,b are $m$ vectors trace $(A\ a\ b^T) = b^T A\ a$, we may rewrite (5.13) as

$$0 = \int_0^T \int_0^T \text{trace } (K^A(t,\sigma,u)(\delta_1 u(t)\ \delta_2 u(\sigma)^T)^T)d\sigma\ dt$$

The constraints (5.5) and the fact that $\delta_1 u$ are piecewise constant imply that $\delta_1 u \otimes \delta_2 u \in D_1 \cap S_1^\perp$. It follows that (5.13) is equivalent to

$$\langle K^A, D_1 \cap S_1^\perp \rangle = 0.$$

However since $D \cap S_1^\perp$ is dense in $S_1^\perp$ we conclude that $K^A \in S_1$. Now $S_1$ may be decomposed into the direct sum $S_1 = S_1^A \otimes S_1^S$ where $S_1^A \cap S_1^S = 0$, and $S_1^A$ is the space of matrix valued functions.

(5.14)     $K_1(t)H(\sigma,u) - H(t,u)^T K_1(\sigma)^T$

for some matrix valued function $K_1$; and $S_1^S$ is the space of matrix valued functions

$$K_2(t)\ H(\sigma,u) + H\ (t,u)^T K_2(\sigma)^T,$$

for some matrix valued function $K_2$. $S_1^A$ consists of those elements K of $S_1$ satisfying $0 = K(t,\sigma) + K(\sigma,t)^T$, whereas $S_1^S$ consists of those elements K of $S_1$ satisfying $0 = K(t,\sigma) - K(\sigma,t)^T$. Clearly, since $K^A \in S_1$ we also have $K^A \in S_1^A$, and so the representation (5.11) follows from (5.14).                    []

Before continuing our series of results we introduce some more notation. Let $P_t^-(u) = P_t^-$ be the projection onto the range of

$$\int_0^t H(\sigma,u) \, H(\sigma,u)^T d\sigma$$

By standard arguments, Brockett [B3], we may partition $[0,\infty)$ into a union $\bigcup_{i=1}^{n-1} (t_i, t_{i+1}] \cup (t_n, \infty)$ such that $P_t^-$ is constant on each subinterval. Since we are dealing with analytic systems, and $H(\cdot,u)$ is piecewise analytic, each $t_i$ coincides with a discontinuity of $u$. (However we shall not make use of this fact.) Note that $P_t^- P_s^- = P_s^- P_t^- = P_s^-$ for $s \le t$. We may also write $P_t^- = \Pi_t \Pi_t^T$, where $\Pi_t = \Pi_t(u)$ is an $k \times k_1(t)$ matrix satisfying $\Pi_t^T \Pi_t = I_{k_1(t)}$, the $k_1(t) \times k_1(t)$ identity matrix, and $k_1(t) = \dim$ range $P_t^-$.

Similarly we may define $P_t^s$, $s \ge t$, as the projection onto the range of

$$\int_t^s H(\sigma,u) \, H(\sigma,u)^T d\sigma.$$

$s \longrightarrow P_t^s$ is piecewise constant as before so $P_t^+(u) = P_t^+ = \sup P_t^s$ is a well defined projection for each $t$. Moreover $t \longrightarrow P_t^+$ is piecewise constant with $P_t^+ P_s^+ = P_s^+ P_t^+ = P_s^+$ for $s \ge t$. We write $P_t^+ = \Sigma_t \Sigma_t^T$, where $\Sigma_t = \Sigma_t(u)$ is an $k \times k_2(t)$ matrix satisfying $\Sigma_t^T \Sigma_t = I_{k_2}(t)$, the $k_2(t) \times k_2(t)$ identity matrix, and $k_2(t) = \dim$ range $P_t^+$. Note also that $P_t^- H(\sigma,u) = H(\sigma,u)$ for $t \ge \sigma$ and $P_t^+ H(\sigma,u) = H(\sigma,u)$ for $t \le \sigma$. Moreover both of the following matrices are invertible

(5.15) $$\Pi_t^T \left( \int_0^t H(\sigma,u) H(\sigma,u)^T d\sigma \right) \Pi_t, \quad \Sigma_t^T \left( \int_t^{s(t)} H(\sigma,u) H(\sigma,u)^T d\sigma \right) \Sigma_t,$$

where $s(t)$ is any finite time satisfying $P_t^{s(t)} = P_t^+$.


LEMMA 5.6

Consider a quasi-minimal, analytic and complete system (5.1). Then for any $T > 0$ and piecewise constant control $u$, there exists another piecewise constant control $\bar{u}$ such that $\bar{u}(t) = u(t)$, $t \in [0,T]$ and $P_T^+(\bar{u})$ is the $k \times k$ identity matrix. Moreover given any $T > 0$ then there exists a piecewise constant control $\bar{u}$ such that $P_T^-(\bar{u})$ is the $k \times k$ identity matrix.

Proof The proof of the first assertion is almost identical to the second so we only consider the second. Recall that each column of $H(t,u)$ represents a vector field of the form $(\psi_{0,t}^u)_* \, g_j(\psi_{t,0}^u(x_0))$. By the strong accessibility assumption the distribution $\mathbf{L}_0(x)$ coincides with $T_x M$ for all $x \in M$. Suppose that we are given s > 0 and a control u on [0,s] and that range $H(t,\bar{u})$ is contained in a fixed proper subspace of $R^k$, as t ranges over an interval $[s,\epsilon)$, $\epsilon > s$, and $\bar{u}$ ranges over all piecewise constant controls with $\bar{u}(t) = u(t)$ for $t \in [0,s]$. By differentiation, and the definition of $L_0$, it is clear that we would then have $L_0(\psi_{s,0}^u(x_0))$ contained in a proper subspace of $T_{\psi_{s,0}^u(x_0)}M$. This contradiction yields a result which we now make use of as follows. Let $[0,T) = \bigcup_{i=1}^{k} [t_i,t_{i+1})$ be a partition of $[0,T)$ into k non empty sub-intervals. Choose a control $u_1$ on $[0,t_2) = [t_1,t_2)$ such that $H(\sigma,u_1) \neq 0$ on $[0,t_2)$. Let $a_1^1,\cdots,a_1^{r_1}$ be a maximal independent set of vectors such that $(a_1^1)^T H(\sigma,u_1) \equiv 0$ on $[0,t_2)$ and let $V_1 = $ span $\{a_1^1,\cdots,a_1^{r_1}\}$. If $V_1 = \{0\}$ we finish. If not, by the above result there exists a control $u_2$ on $[0,t_3)$ with $u_2(t)=u_1(t)$ for $t \in [0,t_2)$, such that range $H(\sigma,u_2) \not\subset V_1^\perp$, for some $\sigma \in [t_2, t_3)$. Let $a_2^1 \cdots a_2^{r_2}$ be a maximal independent set of vectors such that $(a_2^1)^T H(\sigma,u_2) \equiv 0$ on $[t_2, t_3)$, and let $V_2 = $ span $\{a_2^1 \cdots a_2^{r_2}\}$. Since $\bigcap_{\sigma \in [t_2,t_3)}$ ker $H(\sigma,u_2)^T = V_2$ we see $V_2^\perp \not\subset V_1^\perp$. If $V_1 \cap V_2 = \{0\}$ we finish. By repeating the argument above with $V_1^\perp$ replaced by $V_1^\perp + V_2^\perp$, and again repeating the argument we eventually obtain $R^k = V_1^\perp + V_2^\perp + \cdots + V_N^\perp$ and a control $\bar{u}$ on $[0,t_{N+1})$ satisfying $\bigcap_{\sigma \in [t_i,t_{i+1})}$ ker $H(\sigma,\bar{u})^T = V_i$. If $N \neq k$ we define $\bar{u}$ on $[0,T)$, by arbitrarily extending $\bar{u}$ with a piecewise constant control.

Finally, if $v \in R^k$ satisfies $v^T (\int_0^T H(\sigma,\bar{u}) H(\sigma,\bar{u})^T d\sigma) v = 0$ then $v^T H(\sigma,\bar{u}) \equiv 0$ on $[0,T]$, so $v \in V_i$, $i = 1,..,N$. Since

$$\bigcap_{i=1}^{N} V_i = (V_1^\perp + \cdots + V_N^\perp)^\perp = (R^k)^\perp = \{0\}.$$

it follows that $v = 0$ and hence the symmetric matrix $\int_0^T H(\sigma,\bar{u}) H(\sigma,\bar{u})^T d\sigma$ is nonsingular. Inparticular $P_T^-(\bar{u})$ is the $k \times k$ identity matrix as claimed. []

## PROPOSITION 5.7

Consider the situation described in proposition (5.5). Given any T > 0 there exists a piecewise constant control $\bar{u}$ on $[0,T]$ such that for any piecewise constant control u satisfying $u(t) = \bar{u}(t)$, $t \in [0,T]$ we have

(5.16)     $W_v(t,\sigma,u) + W_v(\sigma,t,u)^T = 0 \qquad t,\sigma \geq T.$

Proof Applying proposition (5.5) we see that $K^A(\cdot,\cdot,u)$ satisfies the equation (5.11) for some matrix valued function $\tilde{G}(\cdot,u)$, and every piecewise constant control u. From the definition of $K^A$ we may assume that $\tilde{G}(t,u)$ is independent of the values of the control $u(s)$, for $s > t$. Moreover it is clear that we may write

(5.17)     $\tilde{G}(t,u) = G(t,u) + R_1(t,u)$

(5.18)     $\tilde{G}(t,u) = -G(t,u) + R_2(t,u)$

where

(5.19)     $R_1(t,u)H(\sigma,u) - H(t,u)^T R_2(\sigma,u)^T = 0, \qquad\qquad t > \sigma$

Cleary $R_i(t,u)$, $i = 1,2$, are also independent of the values of the control $u(s)$ for $s > t$. From (5.17) and (5.18) we obtain

(5.20)     $G(t,u) = \frac{1}{2}(R_2(t,u) - R_1(t,u)).$

We may rewrite (5.19) as

(5.21)     $R_1(t,u)P_r^- H(\sigma,u) - H(t,u)^T P_s^+ R_2(\sigma,u)^T = 0, \ t \geq s > r \geq \sigma.$

Writing $P_r^- = \Pi_r \Pi_r^T$, $P_s^+ = \Sigma_s \Sigma_s^T$, by using (5.15), integration, and a technique of Brockett [B3], we obtain

(5.22)     $R_1(t,u)\Pi_r = H(t,u)^T \Sigma_s K_1(s,r), \qquad\qquad t \geq s > r,$

$\Sigma_s^T R_2(\sigma,u)^T = K_2(s,r) \Pi_r^T H(\sigma,u), \qquad\qquad s > r \geq \sigma.$

However substituting these equations into (5.21) we obtain for $t \geq s > r \geq \sigma$

$H(t,u)^T \Sigma_s K_1(s,r)\Pi_r^T H(\sigma,u) = H(t,u)^T \Sigma_s K_2(s,r)\Pi_r^T H(\sigma,u).$

By integration we deduce that $K_1(s,r) = K_2(s,r)$. Thus

(5.23)     $R_2(\sigma,u)\Sigma_s = H(\sigma,u)^T \Pi_r K_1(s,r)^T, \qquad\qquad s > r \geq \sigma.$

From (5.22) we again deduce by integration

$$K_1(s,r) = A_1(s) \, \Pi_r$$

for some matrix valued function $A_1$. Substituting into (5.23) we obtain

$$R_2(\sigma,u) \, \Sigma_s = H(\sigma,u)^T \Pi_r \; \Pi_r^{\ T} A_1(s)^T, \qquad\qquad s > r \geq \sigma$$

Thus again by integration we deduce that

(5.24) $\qquad \Pi_r^{\ T} A_1(s)^T = A_2(r) \, \Sigma_s \qquad , \; s > r$

for some matrix valued function $A_2$. In fact

$$A_2(r) = \left( \int_0^r \Pi_r^{\ T} H(\sigma,u) \; H(\sigma,u)^T \Pi_r \, d\sigma \right)^{-1} \int_0^r \Pi_r^{\ T} H(\sigma,u) R_2(\sigma,u) d\sigma$$

Hence it is clear that $A_2(r)$ is independent of the values of the control $u(s)$ for $s > r$. Recall that $P_s^+(u) = \Sigma_s \, \Sigma_s^T$, and that by lemma (5.6) we may change $u$ to a control $\bar{u}$ which is identical to $u$ on $[0,s]$, but for which $P_s^+(\bar{u})$, and hence $\Sigma_s$ is the identity matrix. We may therefore write (5.24) as

(5.25) $\qquad \Pi_r^{\ T} A_1(s)^T = A_2(r), \qquad\qquad s > r$

since $A_2(r)$ has not been altered by this change in control. We now deduce from (5.25) that there is a matrix $Q = Q(\bar{u})$ such that

$$A_2(r) = \Pi_r^{\ T} Q$$

Now $\Sigma_s K_1(s,r) = \Sigma_s A_1(s) \Pi_r$ so by (5.25) we get $\Sigma_s K_1(s,r) = A_2(r)^T = Q^T \Pi_r$. Applying this result to (5.22) we obtain

(5.26) $\qquad R_1(t,u)\Pi_r = H(t,u)^T Q^T \Pi_r, \qquad\qquad t > r.$

On the other hand $\Pi_r K_1(s,r)^T = \Pi_r \Pi_r^{\ T} A_1(s)^T$ so by (5.24) we get $\Pi_r K_1(s,r)^T = \Pi_r A_2(r) \Sigma_s = Q \Sigma_s$. Applying this to (5.23) we obtain

$$R_2(\sigma,u)\Sigma_s = H(\sigma,u)^T Q \, \Sigma_s, \qquad\qquad s > \sigma.$$

In this case an argument as above gives

(5.27) $\qquad R_2(t,u) = H(t,u)^T Q.$

In any case we may use (5.26) and (5.27) in (5.20) to obtain

(5.28) $\qquad G(t,u)\Pi_r = H(t,u)^T \frac{1}{2} (Q-Q^T) \Pi_r, \qquad t > r.$

We now apply the last part of lemma (5.6), to yield the existence of a control $\bar{u}$ on any interval $[0,T]$ such that $P_t^-(\bar{u}) = \Pi_T \Pi_T^{\ T}$ is the identity matrix. Thus for any control u, which coincides on $[0,T]$ with $\bar{u}$, (5.28) yields

$$G(t,u) = H(t,u)^T \frac{1}{2} (Q-Q^T), \qquad t \geq T.$$

Since

$$W_v(t,\sigma,u) + W(\sigma,t,u)^T = G(t,u) H(\sigma,u) + H(t,u)^T G(\sigma,u)^T$$

$$= H(t,u)^T \frac{1}{2} (Q-Q^T) H(\sigma,u) + H(t,u)^T \frac{1}{2} (Q-Q^T)^T H(\sigma,u) = 0$$

as long as $t,\sigma \geq T$, we have proved (5.16). $\qquad\qquad\qquad\qquad$ []

We note that proposition (5.7) does not quite guarantee that under the conditions of proposition (5.5), every variational system is self adjoint. However as the result below shows, it does guarantee that the variational systems are self adjoint along "periodic" trajectories.

COROLLARY 5.8

Under the conditions of proposition (5.5), for any piecewise constant control u satisfying

$$H(\sigma,u) = H(\sigma+T,u), \ G(t,u) = G(t+T,u), \ t, \ \sigma \geq 0$$

where T is a positive constant depending on u,

$$W(t,\sigma,u) + W(\sigma,t,u)^T = 0, \ t,\sigma \geq 0.$$

Proof Clearly we need only prove the desired identity for $t,\sigma \in [0,T]$. Now for such t and $\sigma$

$$W(t,\sigma,u) + W(\sigma,t,u)^T = G(t,u) H(\sigma,u) + H(t,u)^T G(\sigma,u)^T$$

$$= G(t+T,u) H(\sigma,u) + H(t,u)^T G(\sigma+T,u)^T$$

$$= G(t+T,u)\Pi_\sigma \Pi_\sigma^{\ T} H(\sigma,u) + H(t,u)^T \Pi_t \Pi_t^{\ T} G(\sigma+T,u)^T.$$

Since $t + T > \sigma$ and $\sigma + T > t$, (5.28) yields

$$H(t+T,u)^T \frac{1}{2} (Q-Q^T) H(\sigma,u) + H(t,u)^T \frac{1}{2} (Q-Q^T)^T H(\sigma+T,u)$$

$$= H(t,u)^T \frac{1}{2} (Q-Q^T) H(\sigma,u) + H(t,u)^T \frac{1}{2} (Q-Q^T)^T H(\sigma,u)$$

where we have used "periodicity" again. The result now follows trivially. $\square$

## SECTION 5.3

In this section we present our main results by combining the results of this and the previous sections.

## THEOREM 5.9

Consider a minimal, analytic and complete system (5.1). The system is (locally) Hamiltonian if and only if given any $(u,y) \in \Sigma_e^+ (0)(x_0)$, and admissible variations $(\delta_1 u, \delta_1 y)$ of $(u,y)$ with compact support, $i = 1,2$, we have

$$\int_0^\infty (\delta_2 y(t)^T \delta_1 u(t) - \delta_1 y(t)^T \delta_2 u(t))dt = 0$$

Proof Necessity follows directly from lemma (5.4). To prove sufficiency we apply proposition (5.7) to obtain $T > 0$ and a control $\bar{u}$ on $[0,T]$ such that for any control $u$ coinciding with $\bar{u}$ on $[0,T]$ we have (5.16) i.e.

$$W_v(t,\sigma,u) + W_v(\sigma,t,u)^T = 0, \qquad t,\sigma \geq T.$$

It therefore follows that if system (5.1) is now initialized at $\psi_{T,0}^{\bar{u}}(x_0)$ every variational system is self adjoint. We may therefore apply theorem (4.2), to see that the system is indeed Hamiltonian. $\square$

Before we give the main result stated in chapter (1) we prove the following corollary of theorem (4.2):

## COROLLARY 5.10

If $\Sigma$ is an analytic complete system (5.1) with $g_0(x_0) \in L_0(x_0)$, such that every variational system is self adjoint, then $\Sigma_e^+(0)(x_0)$ may also be realized by a minimal (locally) Hamiltonian system.

<u>Proof</u> Take a minimal realization of $\Sigma_e^+(0)(x_0)$ as guaranteed in Sussmann [S1], denoted $\bar{\Sigma}$. Since the condition that every variational system is self adjoint is a property only of $\Sigma_e^+(0)(x_0)$ it is also true of $\bar{\Sigma}_e^+(0)(x_0)$. Since $\Sigma$ satisfies $g_0(x_0) \in L_0(x_0)$, $\bar{\Sigma}$ is also strongly accessible. Thus by theorem (4.2), $\bar{\Sigma}$ is (locally) Hamiltonian. □

The significance of this result lies in the fact that the system $\bar{\Sigma}$ is indeed minimal, not just quasi-minimal. In the previous works by Van der Schaft [V1,V4] and Goncalves [G1] only quasi-minimal Hamiltonian systems are constructed. It should be pointed out however that the system $\bar{\Sigma}$ will not be globally Hamiltonian in general, only (locally) Hamiltonian (i.e., the internal energy $H_0$ is only locally defined). Using a method in Crouch [C4] it is easy to establish that any minimal (locally) Hamiltonian realization of an input-output map has a quasi-minimal globally Hamiltonian realization.

Having established the existence of minimal Hamiltonian realizations of input-output maps we may now give our main realizability result.

<u>THEOREM 5.11</u>

If $\Phi_\Sigma(x_0)$ is the input-output map of an analytic complete system (5.1) with $g_0(x_0) \in L_0(x_0)$, then $\Phi_\Sigma(x_0)$ has a minimal, analytic and complete Hamiltonian realization, if and only if for any $(\bar{u},\bar{y}) \in \Sigma_e^+(0)(x_0)$, and any two admissible variations $(\delta_i u, \delta_i y)$ of compact support, $i = 1,2,$

$$\int_0^\infty (\delta_2 y(t)^T \delta_1 u(t) - \delta_1 y(t)^T \delta_2 u(t))\, dt = 0.$$

<u>Proof</u> Necessity follows from lemma (5.4) as in theorem (5.9). To prove sufficiency we construct a minimal strongly accessible realization $\bar{\Sigma}$ of $\Phi_\Sigma(x_0)$ as in corollary (5.10). The conditions on $\Sigma_e^+(0)(x_0)$ also hold for $\bar{\Sigma}_e^+(0)(z_0)$ where $z_0 \in \bar{M}$ is the initial state of the minimal realization $\bar{\Sigma}$ on a state space $\bar{M}$. As in theorem (5.9) we conclude that $\bar{\Sigma}$ is Hamiltonian. □

<u>SECTION 5.4</u>

In this section we briefly review the foregoing results of this chapter in the context of non-initialized systems, and show that the external behaviour sets of minimal Hamiltonian systems are characterized formally as Lagrangian submanifolds of the manifold consisting of all external behaviours. We refer to the terminology of subsection (1.6). Our first task is to give a definition of admissible variations which generalizes definition (5.1). We shall for the present revert to the distinction between a variation of a control (behaviour) and a variational field along a control (behaviour).

We consider the following class of non-initialized, analytic and complete systems

(5.28)
$$\dot{x} = g_0(x) + \sum_{i=1}^{m} u_i \, g_i(x), \qquad x \in M$$

$$y_i = H_i(x), \quad 1 \leq i \leq m, \qquad u \in \Omega \subset R^m.$$

Recalling the definitions $\Sigma_i$, $\Sigma_e$, $\Sigma_i^+(T)(x_T)$, $\Sigma_e^+(T)(x_T)$ and variations of elements of these sets, we introduce the sets $\Sigma_i^-(T)(x_T)$, $\Sigma_e^-(T)(x_T)$, by first defining $\Sigma_i^-(T)$ as the restriction to $(-\infty, T]$ of all elements in $\Sigma_i$, and then defining $\Sigma_i^-(T)(x_T)$ as the subset of $\Sigma_i^-(T)$ corresponding to elements $(u,y,x)$ satisfying $x(t) = x_T$. $\Sigma_e^-(T)$ and $\Sigma_e^-(T)(x_T)$ are defined by just projecting out the state trajectory component of each element.

Now we define the sets $\Sigma_i(T_1,T_2)\,(x_{T_1},x_{T_2})(\bar{u},\bar{y},\bar{x})$ for any $(\bar{u},\bar{y},\bar{x}) \in \Sigma_i$, $-\infty < T_1 \leq T_2 < \infty$, and states $x_{T_1} = \bar{x}(T_1)$, $x_{T_2} = \bar{x}(T_2)$ as follows;
$(u,y,x) \in \Sigma_i(T_1,T_2)(x_{T_1},x_{T_2})(\bar{u},\bar{y},\bar{x})$ if

(i)  $(u(t),y(t),x(t)) = (\bar{u}(t),\bar{y}(t),\bar{x}(t))$, $t \in [T_1, T_2)$,
(ii) there exists $(u^-, y^-, x^-) \in \Sigma_i^-(T_1)$ such that $(u,y,x)$ restricted to $(-\infty,T_1)$ coincides with $(u^-, y^-, x^-)$.
(iii) there exists $(u^+, y^+, x^+) \in \Sigma_i^+(T_2)$ such that $(u,y,x)$ restricted to $[T_2,\infty)$ coincides with $(u^+,y^+,x^+)$.

Note that by construction we have $(u,y,x) \in \Sigma_i$ also. We may define $\Sigma_e(T_1,T_2)(x_{T_1},x_{T_2})(\bar{u},\bar{y})$ by projection once more. However for our purpose this definition is not satisfactory since it proposes to describe external behaviour using internal structure. (This is also true for $\Sigma_e^+(T)(x_0)$, but for initialized systems this is not so serious.) We therefore define sets $\Sigma_e(T_1,T_2)(\bar{u},\bar{y})$, for any $(\bar{u},\bar{y}) \in \Sigma_e$, $-\infty < T_1 \leq T_2 < \infty$ as follows; $(u,y) \in \Sigma_e(T_1,T_2)(\bar{u},\bar{y})$ if $(u,y) \in \Sigma_e$ and $(u(t), y(t)) = (\bar{u}(t),\bar{y}(t))$, $t \in [T_1,T_2)$. We define an equivalence relation on $\Sigma_e(T_1,T_2)(\bar{u},\bar{y})$ as follows. $(u,y) \sim (u^*,y^*)$ if given any control $v$ on $[T_1,\infty)$, the pairs $(u_1,y_1)$, $(u_2,y_2) \in \Sigma_e$ defined by

$$u_1(t) = u(t), \; u_2(t) = u^*(t) \quad , \quad t \in (-\infty,T_1)$$

$$u_1(t) = v(t), \; u_2(t) = v(t) \quad , \quad t \in [T_1,\infty),$$

satisfy

$$y_1(t) = y_2(t), \; t \in [T_1,\infty).$$

It is easily verified that if $\Sigma_e$ is the external behaviour of an observable system (5.28), then each equivalence class in $\Sigma_e(T_1,T_2)(\bar{u},\bar{y})$ coincides with a set $\Sigma_e(T_1,T_2)(x_{T_1},x_{T_2})(\bar{u},\bar{y})$ for some choice of $x_{T_1},x_{T_2} \in M$. We are now at liberty to use $\Sigma_e(T_1,T_2)(x_{T_1},x_{T_2})(\bar{u},\bar{y})$ in definitions concerning external behaviour, at least when applied to observable systems. We shall therefore only consider minimal systems in the sequel, proving an analogue of theorem (5.9) but the results could easily be extended, as in theorem (5.11), to strong accessible non observable systems. It is not clear how to extend the results to non strongly accessible systems however, so a true analogue of theorem (5.11) can not be given.

## DEFINITION 5.12

An admissible variation $(u,y)$ of $(\bar{u},\bar{y}) \in \Sigma_e$ is a variation which satisfies the following conditions assuming $(\bar{u},\bar{y})$ is the projection of $(\bar{u},\bar{y},\bar{x}) \in \Sigma_1$:
(i)   $\delta u$ is piecewise constant and supp $\delta u$ is compact
(ii)  supp $\delta y \subset$ supp $\delta u$
(iii) Suppose supp $\delta u \subset [T_1,T_2]$, $\bar{x}(T_1) = x_{T_1}$, $\bar{x}(T_2) = x_{T_2}$ and $(\bar{u}',\bar{y}') \in$

$\Sigma_e(x_{T_1},x_{T_2})(\bar{u},\bar{y})$. Define a map $(t,\epsilon) \to u'(t,\epsilon)$ by setting
$\quad u'(t,\epsilon) = u(t,\epsilon)$ for $t \in [T_1,T_2)$
$\quad u'(t,\epsilon) = \bar{u}'(t)$ for $t \in (-\infty,T_1) \cup [T_2,\infty)$.
We deduce that $u'$ is a variation of $\bar{u}'$, and hence obtain a variation $(u',y')$ of $(\bar{u}',\bar{y}')$. We require the corresponding variational field $(\delta u',\delta y')$ along $(\bar{u}',\bar{y}')$ to satisfy condition (ii) also. Clearly we have $\delta u' = \delta u$.

As before we now abuse notation and say $(\delta u,\delta y)$ is an admissible variation of $(\bar{u},\bar{y}) \in \Sigma_e$ with compact support. Again part of condition (ii) is just causality. The analogue of proposition (5.3) is now stated.

## PROPOSITION 5.13

Consider a minimal, analytic and complete system (5.28). Given any $(u,y) \in \Sigma_e$, where $(u,y)$ is the projection of $(u,y,x) \in \Sigma_1$, and any $T_1,T_2$ with $-\infty < T_1 < T_2 < \infty$, there exists a finite dimensional subspace $S \subset L_2([T_1,T_2]; \mathbb{R}^m)$ and a dense set of piecewise constant functions $\delta u$ in $L_2([T_1,T_2]; \mathbb{R}^m) \cap S^\perp$, which may be realized as a component of an admissible variation $(\delta u,\delta y)$ of $(u,y)$, with compact support contained in $(T_1,T_2)$. Furthermore the corresponding variation $(\delta u,\delta y,\delta x)$ is such that supp $\delta x \subset (T_1,T_2)$.

The proof of this result follows in exactly the same way as proposition (5.3). S may be given as the space of functions $\{H(\cdot,u)^T\alpha;\ \alpha \in \mathbb{R}^n\}$ where $H(t,u)$ is the $k \times m$ matrix whose j-th column is a local coordinate representation of $(\psi^u_{0,t})_*$ $g_j(\psi^u_{t,0}(x(0))$ and the variations $\delta u$ are characterized by

$$(5.29) \qquad 0 = \int_{T_1}^{T_2} H(\sigma,u)\delta u(\sigma)d\sigma$$

As in subsection (5.3) we define $P^s_t$, $s \geq t$, as the projection onto the range of the matrix $\int_t^s H(\sigma,u)\,H(\sigma,u)^T d\sigma$. We also define the related projections

$$(5.30) \qquad P^-_s(u) = \Pi_s\Pi_s^T = \sup_t P^s_t, \qquad P^+_t(u) = \Sigma_t\Sigma_t^T = \sup_s P^s_t$$

Given a minimal system $\Sigma$ described by equations (5.28) ,$(u,y,x) \in \Sigma_i$, and a time T, by an analogue of lemma (5.6) we may find a control $\bar{u}$ such that $\bar{u}(t) = u(t)$, $t \in (-\infty,T)$ (respectively $\bar{u}(t) = u(t)$, $t \in [T,\infty)$) and $P^+_T(\bar{u})$ is the identity matrix (respectively $P^-_T(\bar{u})$ is the identity matrix), and $(\bar{u},\bar{y},\bar{x}) \in \Sigma_i(T,T)(x(T),\ x(T))(u,y,x)$.

Using an analogue of proposition (5.5), which is almost identical in this situation, in an analoguous proof of proposition (5.7) we obtain a more pleasing result for non initialized systems:

PROPOSITION 5.14

Consider a minimal, analytic and complete system (5.28). If for any $(u,y) \in \Sigma_e$, all admissible variations $(\delta_i u,\ \delta_i y)$ of $(u,y)$ with compact support, $i = 1,2$, satisfy

$$\int_{-\infty}^{\infty} (\delta_2 y(t)^T\delta_1 u(t) - \delta_1 y(t)^T\delta_2 u(t))dt = 0,$$

then every variational system is self adjoint.

The precise reason for the better result in this case is that in the proof of proposition (5.7), the improved result in our analogue of lemma (5.6) shows that equations (5.26) and (5.27) may be written as

$$R_1(t,u) = H(t,u)^T Q^T$$

$$R_2(t,u) = H(t,u)^T Q.$$

It follows from (5.20) that if $W_v(t,\sigma,u) = G(t,u)\,H(\sigma,u)$ as usual, then

$$G(t,u) = H(t,u)^T \frac{1}{2} (Q-Q^T).$$

Self adjointness of each variational system, defined simply as

$$(5.31) \qquad W_v(t,\sigma,u) + W_v(\sigma,t,u)^T = 0, \qquad t,\sigma \in R$$

is now a trivial conclusion.

## COROLLARY 5.15

A minimal, analytic and complete system (5.28) is Hamiltonian if and only if the conditions of proposition (5.14) hold.

The proof of this result follows just as its analogue, theorem (5.9), but we need only consider a "subset" $\Sigma_e^+ (0) (x_0)$ of $\Sigma_e$ in the sufficiency proof. It is therefore clear that the assumptions of proposition (5.14) are far stronger than required to merely guarantee that the system is Hamiltonian. We now demonstrate a correspondingly stronger result for non-initialized systems:

## PROPOSITION 5.16

Suppose that (5.28) represents a minimal, analytic, complete and Hamiltonian system $\Sigma$. Let $(u,y) \in \Sigma_e$, and suppose further that $(u,y) \in \overline{\Sigma}_e$ where $\overline{\Sigma}$ is another minimal but not necessarily Hamiltonian system with the same state space M and control constraint set $\Omega$. Let $(Du,Dy)$ be an admissible variation of $(u,y)$ with compact support where $(u,y)$ is viewed as an element of $\overline{\Sigma}_e$. If every admissible variation $(\delta u, \delta y)$ of $(u,y)$ with compact support, where $(u,y)$ is viewed as an element of $\Sigma_e$, satisfies

$$(5.32) \qquad \int_{-\infty}^{\infty} (\delta u(t)^T Dy(t) - \delta y(t)^T Du(t))\, dt = 0,$$

then $(Du,Dy)$ is also an admissible variation of $(u,y)$ viewed as an element of $\Sigma_e$.

## PROOF

With $(u,y) \in \Sigma_e$ we may write

$$(5.33) \qquad \delta y(t) = \int_{-\infty}^{t} W_v(t,\sigma,u)\delta u(\sigma)d\sigma$$

where $W_v(t,\sigma,u)$ satisfies the self adjointness conditions (5.31). Substituting (5.33) into (5.32) we obtain after some manipulation

$$(5.34) \qquad \int_{-\infty}^{\infty} \delta u(t)^T [Dy(t) - \int_t^{\infty} W_v(\sigma,t,u)^T Du(\sigma)d\sigma] \, dt = 0$$

As in proposition (5.13) given any interval $[T_1,T_2]$ we may suppose $\delta u$ takes values in a dense subset of $L_2([T_1,T_2]; R^m) \cap S^{\perp}$ where the constraints $\delta u \in S^{\perp}$ are expressed by equations (5.29), or $0 = \int_{T_1}^{T_2} H(\sigma,u) \, \delta u(\sigma)d\sigma$. Thus from equation (5.34) we deduce that for any $t$

$$(5.35) \qquad Dy(t) - \int_t^{\infty} W_v(\sigma,t,u)^T Du(\sigma)d\sigma = H(t,u)^T a(u)$$

where $a(u)$ is a suitable $k$ vector. We claim $a(u)$ is zero. Suppose that supp $Du \subset [T_1,T_2]$, so by definition supp $Dy \subset [T_1,T_2]$ also. It follows from (5.35) that

$$(5.36) \qquad H(t,u)^T a(u) = 0, \qquad t > T_2$$

Moreover by part (iii) of definition (5.12) of an admissible variation, (5.36) remains valid if we replace $u$ by any control $\bar{u}$ which coincides with $u$ on $(-\infty, T_2)$. Now we may rewrite (5.36) using (5.30) as

$$0 = H(t,\bar{u})^T P_{T_2}^+ (\bar{u}) \, a(\bar{u}) = H(t,\bar{u})^T \Sigma_{T_2} \Sigma_{T_2}^T a(\bar{u}).$$

Hence by integration we see that $0 = \Sigma_{T_2} a(\bar{u})$. However we may choose $\bar{u}$ so that $P_{T_2}^+ (\bar{u})$ and hence $\Sigma_{T_2}$ is the identity matrix. Thus for this $\bar{u}$, $a(\bar{u}) = 0$, and so the left hand side of (5.35) vanishes for this $\bar{u}$. But the left hand side of (5.35) does not depend on $u(t)$ for $t > T_2$ so we see that it vanishes identically, giving

$$(5.37) \qquad Dy(t) = \int_t^{\infty} W_v(\sigma,t,u)^T Du(\sigma)d\sigma.$$

We now use the fact that supp $Dy \subset$ supp $Du \subset [T_1,T_2]$ again to deduce that

$$0 = \int_{T_1}^{\infty} W_v(\sigma,t,u)^T Du(\sigma)d\sigma \text{ for } t < T_1. \text{ Writing } W_v(\sigma,t,u)^T = H(t,u)^T G(\sigma,u)^T \text{ we}$$

obtain

$$0 = H(t,u)^T \int_{T_1}^{\infty} G(\sigma,u)^T Du(\sigma)d\sigma, \qquad t < T_1.$$

But this may be rewritten using (5.30) as

$$0 = H(t,u)^T P_{T_1}^-(u) \int_{T_1}^{\infty} G(\sigma,u)^T Du(\sigma)d\sigma, \qquad t < T_1$$

or

$$(5.38) \qquad 0 = H(t,u)^T \Pi_{T_1} \Pi_{T_1}^T \int_{T_1}^{\infty} G(\sigma,u)^T Du(\sigma)d\sigma, \qquad t < T_1$$

By definition (5.12) both (5.37) and (5.38) are unchanged when u is replaced by any control $\bar{u}$ which coincides with u on $[T_1,\infty)$, and the corresponding element $(\bar{u},\bar{y},\bar{x})$ belongs to $\Sigma_i(T_1,T_2)(x(T_1),x(T_2)(u,y,x)$. We now integrate (5.38) to obtain

$$\Pi_{T_1}^T \int_{T_1}^{\infty} G(\sigma,u)^T Du(\sigma)d\sigma = 0.$$

But we may choose $\bar{u}$, as in our analogue of lemma (5.6), so that $P_{T_1}^-(\bar{u})$ and hence $\Pi_{T_1}$ are the identity matrices. We conclude that

$$\int_{T_1}^{\infty} W_V(\sigma,t,u)^T Du(\sigma)d\sigma = 0.$$

Thus by (5.37) and the fact that supp $Du \subset [T_1,T_2]$

$$Dy(t) = \int_{t}^{\infty} W_V(\sigma,t,u)^T Du(\sigma)d\sigma - \int_{-\infty}^{\infty} W_V(\sigma,t,u)^T Du(\sigma)d\sigma$$

$$= -\int_{-\infty}^{t} W_V(\sigma,t,u)^T Du(\sigma)d\sigma$$

and by the self adjointness condition (5.31)

$$Dy(t) = \int_{-\infty}^{t} W_V(t,\sigma,u)Du(\sigma)d\sigma.$$

This demonstrates that $(Du,Dy)$ is an admissible variation of $(u,y) \in \Sigma_e$ as claimed. $\qquad\qquad\qquad\qquad\qquad\qquad\qquad\qquad\qquad\qquad\qquad\qquad\qquad\qquad$ ☐

As in section (1.6) of chapter (1) we let $N_{M,\Omega,m}$ be the union of all behaviour sets $\Sigma_e$ as $\Sigma$ ranges over all minimal, analytic and complete systems (5.28), and $M_\Sigma^H$ designates a submanifold of $N_{M,\Omega,m}$ consisting of a behaviour set described by a Hamiltonian system $\Sigma$. Using terminology defined in section (1.6) we have our last result of this chapter.

THEOREM 5.17

Each submanifold $M_\Sigma^H$ is a Lagrangian submanifold of $\blacksquare_{M,\Omega,m}$.

Proof  By corollary (5.15) $M_\Sigma^H$ is an isotropic submanifold. By proposition (5.16)

$M_\Sigma^H$ is co-isotropic. Thus $M_\Sigma^H$ is a Lagrangian submanifold.                    □


SECTION 5.5

In this section we shall illustrate the results obtained in this chapter by means
of some very simple examples.
First, as deduced in section 5.1 the admissible variations $(\delta u, \delta y)$ of $(\bar u, \bar y)$ of
compact support on $[0, T]$ are characterized by the condition (5.5)

(5.39)        $\int_0^T H(s,\bar u)\ \delta u(s)ds = 0$

Let us see what this amounts to in the case of the harmonic oscillator (mass-
spring system)

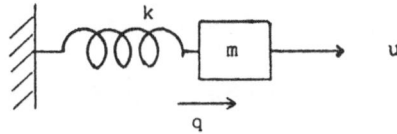

i.e. the linear Hamiltonian input-output system $\dot x = Ax + bu$, $y = cx$ given by

(5.40)     $\dfrac{d}{dt}\begin{bmatrix} q \\ \dot q \end{bmatrix} = \begin{bmatrix} 0 & 1 \\ -\dfrac{k}{m} & 0 \end{bmatrix}\begin{bmatrix} q \\ \dot q \end{bmatrix} + \begin{bmatrix} 0 \\ \dfrac{1}{m} \end{bmatrix} u$ ,        $y = (1\ \ 0)\begin{bmatrix} q \\ \dot q \end{bmatrix}$

Because the system is linear we only have to consider variations $(\delta u, \delta y)$ of
$(\bar u, \bar y) = (0,0)$, which are equal to the actual input-output pairs $(u,y)$. The impulse
response matrix equals $ce^{A(t-s)}b = ce^{At}.e^{-As}b$, and so $H(s,u)$ is given as

(5.41)      $H(s,u) = e^{-As}b = \begin{bmatrix} -\dfrac{1}{m\omega}\sin\omega s \\[2mm] \cos\omega s \end{bmatrix}$

where $\omega = \sqrt{\dfrac{k}{m}}$ is the fundamental frequency of the system. Let us investigate the conditions (5.39) for T equal to the period of the system, i.e., $T = \dfrac{2\pi}{\omega}$,

(5.42) $\qquad \displaystyle\int_{0}^{2\pi/\omega} \begin{bmatrix} -\dfrac{1}{m\omega}\sin\omega s \\[2mm] \cos\omega s \end{bmatrix} u(s)\,ds = 0.$

This yields

(5.43) $\qquad \displaystyle\int_{0}^{2\pi/\omega} u(s)\sin\omega s\,ds = \int_{0}^{2\pi/\omega} u(s)\cos\omega s\,ds = 0$

or, equivalently, the Fourier series of $u(s)$ on $[0,T]$ can be written as

(5.44) $\qquad u(s) = \dfrac{a_0}{2} + \displaystyle\sum_{n=2}^{\infty}(a_n\cos n\omega s + b_n\sin n\omega s)$

Concluding, the admissible variations of $(\bar{u},\bar{y}) = (0,0)$ of compact support on $(0,2\pi/\omega)$ are all those input-output pairs $(u,y)$ for which the input $u$ does not contain the fundamental frequency $\omega = \sqrt{\dfrac{k}{m}}$

As a second example let us consider the mass-spring system with a spring force proportional to $q^3$, i.e.

(5.45) $\qquad m\ddot{q} = -kq^3 + u$

Defining $p = mq$ we obtain, with $H_0 = \dfrac{p^2}{2m} + \dfrac{k}{4}q^4$, $H_1 = q$, the nonlinear Hamiltonian system

(5.46)
$$\dot{q} = \dfrac{\partial H_0}{\partial p} - u\,\dfrac{\partial H_1}{\partial p} = \dfrac{p}{m}$$
$$y = q$$
$$-\dot{p} = \dfrac{\partial H_0}{\partial q} - u\,\dfrac{\partial H_1}{\partial q} = kq^3 - u$$

The variational equations are given as

$$(5.47) \quad \begin{bmatrix} \dot{q}_1 \\ \dot{q}_2 \end{bmatrix} = \begin{bmatrix} 0 & \frac{1}{m} \\ -3k\bar{q}^{-2} & 0 \end{bmatrix} \begin{bmatrix} q_1 \\ q_2 \end{bmatrix} + \begin{bmatrix} 0 \\ 1 \end{bmatrix} u^v$$

$$y^v = (1 \ 0) \begin{bmatrix} q_1 \\ q_2 \end{bmatrix}$$

and the adjoint variational equations as

$$(5.48) \quad \begin{bmatrix} \dot{p}_1 \\ \dot{p}_2 \end{bmatrix} = \begin{bmatrix} 0 & 3k\bar{q}^{-2} \\ -\frac{1}{m} & 0 \end{bmatrix} \begin{bmatrix} p_1 \\ p_2 \end{bmatrix} + \begin{bmatrix} -1 \\ 0 \end{bmatrix} u^a$$

$$y^a = (0 \ 1) \begin{bmatrix} p_1 \\ p_2 \end{bmatrix}$$

Let us consider the (adjoint) variational equations along the equilibrium trajectory $\bar{q} \equiv \bar{p} \equiv \bar{u} \equiv \bar{y} \equiv 0$. Then, since $\bar{q} \equiv 0$, (5.47) and (5.48) easily yield

$$q_2(t) = \int_0^t u^v(\sigma)d\sigma, \quad q_1(t) = \frac{1}{m} \int_0^t \int_0^\sigma u^v(s)ds d\sigma = y^v(t)$$

$$(5.49)$$

$$p_1(t) = -\int_0^t u^a(\sigma)d\sigma, \quad p_2(t) = \frac{1}{m} \int_0^t \int_0^\sigma u^a(s)ds d\sigma = y^a(t)$$

Thus if $u^v \equiv u^a$ we obtain $y^v \equiv y^a$, and so this variational system is self-adjoint (as it should be). A variation $(u^v = \delta u, \ y^v = \delta y)$ has compact support on $(0, T)$ if

$$(5.50) \quad \int_0^T H(\sigma, 0) \ u^v(\sigma) \ d\sigma = 0$$

where $H(\sigma, 0) = (-\frac{\sigma}{m} \ 1)^T$. This yields

$$(5.51a) \quad \int_0^T u^v(\sigma)d\sigma = 0$$

(5.51b) $\qquad \int_0^T \sigma u^v(\sigma) d\sigma = 0$

Using (5.51a) we immediately deduce that (5.51b) is equivalent to

(5.52) $\qquad \int_0^T \int_0^\sigma u^v(s) ds d\sigma = 0$

which is by (5.49) just the condition $y^v(T) = 0$. Now consider two variations $(\delta u^1, \delta y^1)$, $(\delta u^2, \delta y^2)$ of compact support on $(0, T)$. Then

$$m \int_0^\infty [\delta y^2(t) \delta u^1(t) - \delta y^1(t) \delta u^2(t)] dt =$$

$$= \int_0^T [\delta u^1(t) \int_0^t \int_0^\sigma \delta u^2(s) ds d\sigma - \delta u^2(t) \int_0^t \int_0^\sigma \delta u^1(s) ds d\sigma] dt =$$

$$= \int_0^T \frac{d}{dt} [\int_0^t \delta u^1(\sigma) d\sigma \int_0^t \int_0^\sigma \delta u^2(s) ds d\sigma -$$

$$\int_0^t \delta u^2(\sigma) d\sigma \int_0^t \int_0^\sigma \delta u^1(s) ds d\sigma] dt =$$

$$= [\int_0^t \delta u^1(\sigma) d\sigma \int_0^t \int_0^\sigma \delta u^2(s) ds d\sigma - \int_0^t \delta u^2(\sigma) d\sigma \int_0^t \int_0^\sigma \delta u^1(s) ds d\sigma]_0^T =$$

$$= 0, \text{ by (5.51a) or (5.52)},$$

and so Theorem 5.11 is confirmed.

## 6. GENERAL NONLINEAR SYSTEMS

In this chapter we shall briefly show how the self-adjointness and variational criterion for Hamiltonian systems as developed in chapters 4 and 5 for systems (2.1) can be extended to general nonlinear systems

$$\dot{x} = f(x,u) , \qquad x(0) = x_0 \in M$$

(6.1)

$$y_j = h_j(x,u), \qquad j = 1,..,m, \; u = (u_1,..,u_m) \in \Omega$$

Here as before M denotes the k-dimensional state space manifold and $\Omega$ is assumed to be (an open subset of) $\mathbb{R}^m$. We shall not be very precise about the class of admissible controls u(t), as long as they include the piecewise constant right continuous functions. Later on we shall add some __completeness__ assumptions concerning the vectorfields $f(\cdot,u)$ for every $u \in \Omega$. Throughout we shall again assume that all data in (6.1) are (real-)__analytic.__

In order to facilitate our discussions we first recall the more general, geometric interpretation of a nonlinear system (6.1) as introduced in [W1]. Let W be a 2m-dimensional analytic manifold, representing the spoace of external variables (outputs and inputs), with local coordinates $(y_1,\cdots,y_m,u_1,\cdots,u_m)$. Then equations (6.1) can be interpreted as describing a (k+m)-dimensional submanifold L of the product manifold TM×W, which is locally parametrized by the coordinates $x = (x_1,\cdots,x_k)$ for M and m of the coordinates for W, namely $(u_1\cdots,u_m)$:

## DEFINITION 6.1 ([W1],[V1],[V6])

A (general) nonlinear system with state space M and space of external variables W is given by a submanifold $L \subset TM \times W$ locally parametrized by the coordinates of M and some coordinates of W.

Remark Usually it is also required that L is "globally" parametrized in the sense that there exists a __fibre bundle__ B over M (with projection $\pi$) and a mapping $G : B \to TM \; W$, such that G(B) = L, making the diagram

$$B \xrightarrow{\;\;G\;\;} TM \times W$$

$$\pi \searrow \qquad \swarrow \text{projection}$$

$$M$$

commutative. For a discussion of these issues we refer to [B1,W1,V6].

A general nonlinear system is called Hamiltonian if the following holds (cf.[V1,V2]). M has to be a 2n-dimensional symplectic manifold with symplectic form $\omega$, and W also has to be a symplectic manifold with symplectic form $\omega^e$. The symplectic form $\omega$ induces a symplectic form $\dot{\omega}$ on TM ([T,V2]). Let $\omega$ be locally given by $\omega = \sum_{i=1}^{n} dp_i \wedge dq_i$, then locally $\dot{\omega} = \sum_{i=1}^{n} (d\dot{p}_i \wedge dq_i + dp_i \wedge d\dot{q}_i)$. Now we require that $L \subset TM \times W$, describing the system equations (6.1), is a $\underline{\text{Lagrangian}}$ submanifold of $TM \times W$ with its product symplectic form $\dot{\omega} \oplus (-\omega^e)$. Then (cf.[V1,V2]) locally we can take semi-canonical coordinates $(y_1, \cdots, y_m, u_1, \cdots, u_m)$ for W, i.e. $\omega^e = \sum_{i=1}^{n} c_i du_i \wedge dy_i$ with $c_i = \pm 1$, such that L is parametrized by the coordinates x of M and $u = (u_1, \cdots, u_m)$. Furthermore because L is Lagrangian there exists locally a $\underline{\text{generating function}}$ H(x,u) for L. This implies that L as a submanifold of $TM \times W$ is locally given by the equations

$$
\begin{array}{lll}
(6.2) & \dot{x} = X_H(x,u) & x(0) = x_0 \in M \\
& y_j = -c_j \dfrac{\partial H}{\partial u_j}(x,u), & j = 1, \cdots, m, \qquad c_j = \pm 1
\end{array}
$$

where for every u, $X_H(\cdot,u)$ is the locally Hamiltonian vectorfield on M with Hamiltonian function $H(\cdot,u)$.

Comparing (6.2) with (6.1) we conclude that a local coordinate expression (6.1) is Hamiltonian if for every u, $f(\cdot,u)$ is a Hamiltonian vectorfield with Hamiltonian function $H(\cdot,u)$, and if $h_j(x,u)$ equals $\pm \dfrac{\partial H}{\partial u_j}(x,u)$, $j = 1, \cdots, m$.

$\underline{\text{Remark}}$ If the generating function H(x,u) is of the form $H_0(x) - \sum_{i=1}^{m} u_j H_j(x)$, then we recover the definition of an (affine) Hamiltonian system as given in (2.3).

Now we proceed to the definition of the $\underline{\text{prolongation}}$ and $\underline{\text{Hamiltonian extension}}$ of a general nonlinear system. In local coordinates we define the $\underline{\text{variational system}}$ of (6.1) along a certain control u(t) as the time-varying linear system

$$
\begin{array}{lll}
(6.3) & \dot{v}(t) = \dfrac{\partial f}{\partial x}(x(t),u(t))v(t) + \dfrac{\partial f}{\partial u}(x(t),u(t))u^v(t), & v(0) = v_0
\end{array}
$$

$$
y^v(t) = \dfrac{\partial h}{\partial x}(x(t),u(t))v(t) + \dfrac{\partial h}{\partial u}(x(t),u(t))u^v(t)
$$

with $h = (h_1, \cdots, h_m)^T$, and the adjoint system as

(6.4) $\qquad -\dot{p}(t) = (\frac{\partial f}{\partial x})^T (x(t), u(t)) p(t) + (\frac{\partial h}{\partial x})^T (x(t), u(t)) u^a(t), \qquad p(0) = p_0$

$\qquad\qquad y^a(t) = (\frac{\partial f}{\partial u})^T (x(t), u(t)) p(t) + (\frac{\partial h}{\partial u})^T (x(t), u(t)) u^a(t)$

It is easily checked that Lemma 2.1 goes through for this more general definition of the variational and adjoint system. As before we call the original system (6.1) together with (6.3), respectively (6.4), the prolongation, resp. Hamiltonian extension of (6.1). The coordinate-free definitions for a nonlinear system as in Definition 6.1 follow. Since $L \subset TM \times W$ we have $TL \subset T(TM) \times TW$. Applying the canonical involution on $T(TM)$ (in local natural coordinates $(x,v)$ for $TM$ given by $(x,v,\dot{x},\dot{v}) \rightarrow (x,\dot{x},v,\dot{v})$, cf.[T,V1]) we therefore obtain a new nonlinear system with state space $TM$ and space of external variables $TW$. This system is called the prolongation, since if we denote natural coordinates for $TM$ by $(x,v)$ and for $TW$ by $(y,u,y^v,u^v)$ then we recover locally the equations (6.1) together with (6.3). Furthermore it can be seen that in the affine case (2.1) - (3.1) this definition of the prolongation reduces to the global definition of the prolongation given in equation 2.17 of chapter (2).

In general, if $N$ is a submanifold of a product manifold $Q_1 \times Q_2$, then we can associate to $N$, in a canonical way, a submanifold $N_{lift}$ of $T^*Q_1 \times T^*Q_2$, which is a Lagrangian submanifold of $T^*Q_1 \times T^*Q_2$ with the product symplectic form $\omega_1 \oplus (-\omega_2)$ ($\omega_1$, resp. $\omega_2$, denote the natural symplectic forms on $T^*Q_1$, resp. $T^*Q_2$.) Namely, let $\pi_{Q_1}$ and $\pi_{Q_2}$ denote the natural projections of $T^*Q_1$ and $T^*Q_2$ on $Q_1$ and $Q_2$. Then

$$N_{lift} = \{(a_1, a_2) \in T^*Q_1 \times T^*Q_2 | (\pi_{Q_1}(\alpha_1), \pi_{Q_2}(\alpha_2)) = (q_1, q_2) \in N,$$

$$\text{and for all } (X_1, X_2) \in T_{(q_1,q_2)}N, \ \alpha_1(X_1) = \alpha_2(X_2)\}$$

Therefore, since $L \subset TM \times W$ we can define the Lagrangian submanifold $L_{lift}$ of $T^*(TM) \times T^*M$. Because $T^*(TM)$ with its natural symplectic form as a cotangent bundle is symplectomorphic to $T(T^*M)$ with the symplectic form $\dot{\Omega}$ induced by the natural symplectic form $\Omega$ on $T^*M([T])$, we also have that $L_{lift}$ is a Lagrangian submanifold of $T(T^*M) \times T^*W$ with its product symplectic form $\dot{\Omega} \oplus (-\Omega^e)$ (where $\Omega^e$ is the natural symplectic form on $T^*W$). Hence we have obtained a Hamiltonian system with state space $T^*M$ and space of external variables $T^*W$. The generating function of $L_{lift}$ is given by

(6.5) $\qquad H(x,p,u,u^a) = p^T f(x,u) + (u^a)^T h(x,u)$

where $(x,p)$ are natural coordinates for $T^*M$ and $(y,u,-u^a,y^a)$ are natural coordinates for $T^*W$. (Notice the order and the signs of $u^a$ and $y^a$.) This system is called the <u>Hamiltonian extension</u>, since in local coordinates as above we recover the local expressions

$$\dot{x} = \frac{\partial H}{\partial p} = f(x,u)$$

$$\dot{p} = -\frac{\partial H}{\partial q} = -\left(\frac{\partial f}{\partial x}\right)^T(x,u)p - \left(\frac{\partial h}{\partial x}\right)^T(x,u)\ u^a$$

(6.6)

$$y = \frac{\partial H}{\partial u^a} = h(x,u)$$

$$y^a = \frac{\partial H}{\partial u} = \left(\frac{\partial f}{\partial u}\right)^T(x,u)\ p + \left(\frac{\partial h}{\partial u}\right)^T(x,u)\ u^a$$

For a suitable definition of <u>minimality</u> of a general analytic nonlinear system we take recourse to the notion of <u>extended system</u> as defined in [V1,V6]. In local coordinates the extended system is simply obtained by adding an integrator to all the input channels of (6.1)

$$\dot{x} = f(x,u) \qquad x(0) = x_0$$

(6.7) $\qquad \dot{u} = v \qquad\qquad u(0) = 0$

$$y = h(x,u)$$

and regarding (6.7) as a system with $(k + m)$-dimensional state $(x,u)$ and m inputs v. Note that the extended system is a system of the form (3.1) and hence we can apply the minimality notions of Section 3. The accessibility algebra $L_{ext}$ of the extended system is the Lie algebra generated by the vectorfields $f(x,u)\frac{\partial}{\partial x}$ and $\frac{\partial}{\partial u_j}$, $j = 1, \cdots, m$, on $(x,u)$, and the strong accessibility algebra $L_{0\ ext}$ is the ideal in $L_{ext}$ generated by $\frac{\partial}{\partial u_1}, \cdots, \frac{\partial}{\partial u_m}$. Now we <u>define</u> (6.1) to be <u>strongly accessible</u> if its extended system is (strongly)accessible, or equivalently if $\dim L_{ext}(x,u) = k + m$ or $\dim L_{0\ ext}(x,u) = k + m$ for all $(x,u)$. (In order to show the consistency of this definition one has to show that for a nonlinear system already of the form (3.1) strong accessibility is equivalent to strong accessibility of its extended system. This is done in [V1,V6].) Furthermore the <u>observation space</u> $H_{ext}$ of (6.7) is given as the linear space of functions on $(x,u)$ of the form $L_{f_1} \cdots L_{f_s} h_j$, with $f_r$, $r = 1, \ldots, s$, equal to $f(x,u)\frac{\partial}{\partial x}$ or $\frac{\partial}{\partial u_i}$, $i=1,\ldots,m$, and $j=1,\ldots,m$. In order to define observability of a nonlinear system (6.1) we have to add an extra set of outputs $w_j$ to the extended system (6.7)

(6.8)        $w_j = u_j$,     $j = 1 \cdots, m$

Clearly the observation space of (6.7) together with (6.8) is $H_{ext}$ plus the functions $u_1, \ldots, u_m$. Now we _define_ (6.1) to be (_weakly_)_observable_ if (6.7) together with (6.8) is (weakly) observable. Equivalently, $H_{ext}$ together with the functions $u_1, \ldots, u_m$ has to distinguish (nearby) points in $(x, u)$. (The consistency of this definition is checked in [V1].) Because of analyticity it again ([H]) follows that if (6.1) is accessible then the dimension of the codistribution $d_x H_{ext}$ $(x, u)$ is constant for each $(x, u)$, so that (6.1) is weakly observable if and only if dim $d_x H_{ext}$ = dim M. ($d_x$ denotes differentiations with respect to the x-coordinates.) Finally (6.1) is called (quasi-) _minimal_ if it is strongly accessible and (weakly) observable. Since the _prolongation_ of an extended system is equal to the extended system taken of the prolongation of the system, it immediately follows that Corollary 3.3. holds for general nonlinear systems as well: (6.1) is (quasi-) minimal if and only if its prolongation is (quasi-) minimal. With respect to the generalization of Proposition 3.4 and Corollary 3.7 we make the following observations. The observation space $H_{ext}$ for a general Hamiltonian system (6.2) is spanned by the functions $\frac{\partial H}{\partial u_1}, \cdots, \frac{\partial H}{\partial u_m}$, together with all (repeated) Poisson brackets (with respect to the Poisson bracket on M) of these functions with $H(x, u)$ and all (repeated) differentiations to $u_1, \ldots, u_m$ ([V1]). The general Hamiltonian system (6.2) is _quasi-minimal_ if and only if the dimension of the codistribution $d_x H_{ext}$ is equal to the dimension of M, and _minimal_ if furthermore $H_{ext}$ together with the functions $u_1, \ldots, u_m$ distinguishes points in $(x, u)$ (not only nearby points as in the quasi-minimal case). The observation space $H_{ext}^e$ of the Hamiltonian _extension_ of a general nonlinear system (6.1) is given as follows. Since the Hamiltonian is of the form $H(x, p, u, u^a) = p^T f(x, u) + (u^a)^T h(x, u)$ we have

$$\frac{\partial H}{\partial u_j^a} = h_j(x, u)$$

(6.9)                                              $j = 1, \ldots, m$

$$\frac{\partial H}{\partial u_j} = (\frac{\partial f}{\partial u_j})^T (x, u) p + (\frac{\partial h}{\partial u_j})^T (x, u) u^a = y_j^a$$

Furthermore similarly to Lemma 3.6

$$(6.10) \quad \{H(x,p,u,u^a), \frac{\partial H}{\partial u_j^a}\} = L_f h_j$$

$$\{H(x,p,u,u^a), \frac{\partial H}{\partial u_j}\} = p^T[f(x,u)\frac{\partial}{\partial x}, \frac{\partial f}{\partial u_j}(x,u)\frac{\partial}{\partial x}]$$

As in Theorem 3.5 we therefore obtain that $H_{ext}^e = p^T L_{0\,ext} + H_{ext}^\ell$, and as in Corollary 3.7 we have that (6.1) is (quasi-) minimal if and only if its Hamiltonian extension is (quasi-) minimal.


As in Definition 3.1 we call a variational system (6.3) <u>self-adjoint</u> if its input-output map for $v_0 = 0$ is equal to the input-output map of the adjoint system (6.4) for $p_0 = 0$. (As before we will assume throughout that the vectorfields $f(\cdot,u)$ for all constant $u \in \Omega$ are <u>complete</u>.) In local coordinates the input-output map of the variational system is of the form

$$(6.11) \quad y^v(t) = \int_0^t W_v(t,\sigma,u)u^v(\sigma)d\sigma + \frac{\partial h}{\partial u}(x(t),u(t))u^v(t)$$

whereas the input-output map of the adjoint system is given by

$$(6.12) \quad y^a(t) = \int_0^t W_a(t,\sigma,u)u^a(\sigma)d\sigma + (\frac{\partial h}{\partial u})^T(x(t),u(t))u^a(t)$$

where $W_a(t,\sigma,u) = -W_v^T(\sigma,t,u)$ (cf.(4.11)).

Hence a variational system along a control $u$ is self-adjoint if and only if (compare with (4.12))

$$(6.13) \quad W_v(t,\sigma,u) = -W_v^T(\sigma,t,u) \qquad\qquad t,\sigma \geq 0$$

$$\frac{\partial h}{\partial u}(x(t),u(t)) = (\frac{\partial h}{\partial u})^T(x(t),u(t)) \qquad t \geq 0$$

Before proceeding to the generalisation of Theorem 4.2 for general nonlinear systems we have to remark on the Sussmann uniqueness theorem which is crucially used in the proof of that theorem (cf. Lemma 4.3). In the Sussmann theorem it is assumed that the input and output space of the system are <u>globally</u> defined. In the terminology of Definition 6.1 this means that $W = Y \times U$, with $Y$, resp. $U$, the output, resp. input manifold, and that $L \subset TM \times W$ is globally parametrized by $M \times U$ (so that $f(\cdot,u)$ are globally defined vectorfields for each $u \in U$). Furthermore it is assumed that $Y = \mathbb{R}^m$ and that $U$ is (a part of) $\mathbb{R}^m$. In order to avoid complications we shall therefrom <u>assume</u> throughout that the nonlinear systems are such that $W = Y \times U = \mathbb{R}^m \times \mathbb{R}^m$, and that $f(\cdot,u)$ is a well-defined vectorfield for each $u \in \mathbb{R}^m$. Since in the Sussmann theorem the outputs $y$ do not directly depend on the inputs u we still need a generalisation of that theorem. Recall that a system is called complete if the associated vectorfields for constant input are complete.

PROPOSITION 6.2

Consider two minimal systems $\Sigma_1$, $\Sigma_2$

$$\Sigma_1: \begin{array}{l} \dot{x}^1 = f^1(x^1, u^1) \\ y^1 = h^1(x^1, u^1) \end{array} , \; x \in M_1 \quad \Sigma_2: \begin{array}{l} \dot{x}^2 = f^2(x^2, u^2) \\ y^2 = h^2(x^2, u^2) \end{array} , \; x \in M_2$$

such that the extended systems are complete. Assume that the input-output maps of the systems initialized in $x_1 \in M_1$ and $x_2 \in M_2$ are equal. Then there exists a unique diffeomorphism $\phi: M_1 \to M_2$ with $\phi(x_1) = x_2$ such that

i) $\qquad \phi_* f^1(\cdot, u) = f^2(\cdot, u)$, for all $u \in \mathbb{R}^m$

ii) $\qquad h^2(\phi(x), u) = h^1(x, u)$, for all $u \in \mathbb{R}^m$, $x \in M_1$

Proof The trick is to consider the extended systems of $\Sigma_1$ and $\Sigma_2$. It is clear that the input-output maps of the extended systems (6.7) together with (6.8) initialized in $(x_1, 0)$, respectively $(x_2, 0)$, are equal if and only if the input-output maps of $\Sigma_1$ and $\Sigma_2$ initialized in $x_1$, resp. $x_2$, are equal. Hence by the Sussmann theorem for systems (6.7) together with (6.8) there exists a unique diffeomorphism $\psi: M_1 \times \mathbb{R}^m \to M_2 \times \mathbb{R}^m$ with $\psi(x_1, 0) = (x_2, 0)$ such that

i) $\qquad \psi_* f^1 = f^2$

ii) $\qquad \psi_* \dfrac{\partial}{\partial u_j} = \dfrac{\partial}{\partial u_j} \qquad j = 1, \cdots, m$

iii) $\qquad h^2 \circ \psi = h^1$

where $(u_1, \cdots, u_m)$ are the standard coordinates for $U = \mathbb{R}^m$. By ii) it follows that $\psi$ is of the form $\psi(x, u) = (\phi(x), \alpha(x, u))$ for certain maps $\phi: M_1 \to M_2$ and $\alpha: M_1 \; U \to U$. Using i) and ii) we have

$$\psi_* \frac{\partial f^1}{\partial u_j} \frac{\partial}{\partial x} = \psi_* [f^1, \frac{\partial}{\partial u_j}] = [f^2, \frac{\partial}{\partial u_j}] = \frac{\partial f^2}{\partial u_j} \frac{\partial}{\partial x}$$

and in general $\psi_* L^1_{0 \; ext} = L^2_{0 \; ext}$. Using strong accessibility, the structure of $L_{0 \; ext}$, and the fact that $\psi(x_1, 0) = (x_2, 0)$ it follows that $\alpha(x, u) = u$ for all $x, u$, so that $\psi(x, u) = (\phi(x), u)$. The rest of the proof follows easily. $\qquad \square$

The generalisation of Theorem 4.2 is now straightforward.

## THEOREM 6.3

Let (6.1) be a minimal nonlinear system, such that the extended system is complete. Then the system is a Hamiltonian system (6.2) if and only if every variational system is self-adjoint.

Proof For the easy "only if" direction we refer to the proof of Theorem 4.2. Suppose every variational system is self-adjoint. Then equivalently the input-output map of the prolongation and Hamiltonian extension initialized in $(x_0,0)$ are equal. Hence by Proposition 6.2 there exists a unique diffeomorphism $\phi: TM \rightarrow T^*M$ transforming the prolongation into the Hamiltonian extension. As in Lemma 4.4 and 4.5 it follows that $\phi(x,v) = (x,\omega(x)v)$, with $\omega(x)$ a non-singular anti-symmetric matrix representing a symplectic form $\omega = \sum_{i=1}^{2n} \omega_{ij}(x)dx_i \wedge dx_j$ on M. Furthermore for every u the vectorfields $f(\cdot,u)$ are locally Hamiltonian. Hence there exists locally a function $\tilde{H}: M \times \mathbb{R}^m \rightarrow \mathbb{R}$ such that $f^T(x,u)\omega(x) = -\frac{\partial \tilde{H}}{\partial x}(x,u)$. Since $(\frac{\partial f}{\partial u_j})^T(x,u)\ \omega(x) = \frac{\partial h_j}{\partial x}(x,u)$, $j = 1,\cdots,m$, we also have that $-\frac{\partial}{\partial u_j}(\frac{\partial \tilde{H}}{\partial x}) = \frac{\partial h_j}{\partial x}$, $j = 1,\ldots,m$, or equivalently

$$(6.14) \qquad \frac{\partial}{\partial x_i}(\frac{\partial \tilde{H}}{\partial u_j}) = -\frac{\partial h_j}{\partial x_i} \qquad\qquad j = 1,\cdots,m, \ i = 1,\cdots,2n$$

Hence it follows that there exist functions $k_j(u)$ such that

$$(6.15) \qquad h_j = -\frac{\partial \tilde{H}}{\partial u_j} + k_j(u) \qquad\qquad j = 1,\cdots,m$$

By the second line of (6.13) it follows that $\frac{\partial h_j}{\partial u_i} = \frac{\partial h_i}{\partial u_j}$ and hence

$$(6.16) \qquad \frac{\partial k_j}{\partial u_i} = \frac{\partial h_j}{\partial u_i} + \frac{\partial^2 \tilde{H}}{\partial u_i \partial u_j} = \frac{\partial h_i}{\partial u_j} + \frac{\partial^2 \tilde{H}}{\partial u_j \partial u_i} = \frac{\partial k_i}{\partial u_j} \qquad i,j = 1,\cdots,m$$

So there exists locally a function K(u) with $k_j = \frac{\partial K}{\partial u_j}$.

Then $H(x,u) = \tilde{H}(x,u) + K(u)$ is the Hamiltonian of the system, i.e. for this H the system is given by (6.2). □

Remark Note that the condition of completeness of the extended system means that for every $u \in R^m$ the time-varying vectorfield $\dot{x}(t) = f(x(t),ut)$, $x(0) = x_0$, is complete.

Finally we shall briefly show that also the variational criterion for Hamiltonian systems as developed in chapter (5) goes through for general nonlinear systems. Recall that the variational system along u is given by .

$$(6.17) \qquad \dot{v}(t) = A(t)v(t) + B(t)u^v(t) \qquad\qquad v(0) = v_0$$
$$\phantom{(6.17) \qquad} y^v(t) = C(t)v(t) + D(t)u^v(t)$$

where $A(t) = \frac{\partial f}{\partial x}(x(t),u(t))$, $B(t) = \frac{\partial f}{\partial u}(x(t),u(t))$, $C(t) = \frac{\partial h}{\partial x}(x(t),u(t))$ and

$D(t) = \frac{\partial h}{\partial u}(x(t),u(t))$. Hence the input-output map for $v_0 = 0$ is given by

$$(6.18) \qquad y^v(t) = \int_0^t W_v(t,\sigma,u)\, u^v(\sigma)d\sigma + D(t)u^v(t)$$

where $W_v(t,\sigma,u) = C(t)\,\phi^u(t,\sigma)\,B(\sigma)$ and $\phi^u(t,\sigma)$ is the unique solution of

$$(6.19) \qquad \frac{\partial}{\partial t}\phi^u(t,\sigma) = A(t)\,\phi^u(t,\sigma), \qquad \phi(\sigma,\sigma) = I$$

Because we assume that the vectorfields $f(\cdot,u)$ for constant u are complete, $W_v(t,\sigma,u)$ is actually defined for all $t \geq \sigma \geq 0$ and is in local coordinates given by

$$(6.20) \qquad W_v(t,\sigma,u) = G(t,u)\,H(\sigma,u)$$

with

$$(6.21) \qquad G(t,u) = C(t)\,\phi(t,0), \quad H(\sigma,u) = \phi(\sigma,0)\,B(\sigma)$$

Remark: Let $\bar{u}$ be a constant control and let $\psi_{t,\sigma}(\bar{x},\bar{u})$ denote the flow of the vectorfield

$$\dot{x} = f(x,u) \qquad\qquad x(\sigma) = \bar{x}$$
$$\dot{u} = 0 \qquad\qquad u(\sigma) = \bar{u}$$

on the extended state space $M \times R^m$. Then analogously to (4.9)

$$(6.22) \qquad W_v(t,\sigma,\bar{u})_{ij} = dh_i(\psi_{t,0}(x_0,\bar{u}))(\psi_{t,\sigma})_*[f, \frac{\partial}{\partial u_j}](\psi_{\sigma,0}(x_0,\bar{u}))$$

where $dh_1$ is a one-form on $M \times \mathbb{R}^m$, and $[f, \frac{\partial}{\partial u_j}]$ denotes the Lie bracket of the vectorfields $f(x,u) \frac{\partial}{\partial x}$ and $\frac{\partial}{\partial u_j}$ on $M \times \mathbb{R}^m$. For _piecewise_ constant controls $u$ a similar expression is obtained by concatenation.

As in definition 5.1 a variation $(\delta u, \delta y)$ of $(u,y) \in \Sigma_e(0)(x_0)$ is _admissible_ if $\delta u$ is piecewise constant, supp $\delta y \subset$ supp $\delta u \subset (0,T)$ for some finite $T$, and also condition (iii) of definition 5.1 holds. As in proposition 5.3 these variations are constrained only by the conditions

$$(6.23) \qquad \int_0^T H(\sigma,u)\delta u(\sigma)d\sigma = 0$$

Moreover as in proposition (3.8) we may prove that checking (weak) observability of the prolonged system of a general nonlinear system may be done by setting $u^V$ ($=\delta u$) equal to zero. It is easily seen ([V1]) that equation (4.36) remains true for general Hamiltonian systems. All this yields the immediate generalization of Lemma 5.4

## LEMMA 6.4

Consider a quasi-minimal, analytic, complete, general Hamiltonian system (6.2). Given any $(u,y) \in \Sigma_e^+(0)(x_0)$, and admissible variations $(\delta_i u, \delta_i y)$ of $(u,y)$ with compact support, $i = 1,2$, we have

$$(6.24) \qquad \int_0^\infty (\delta_2^T y(t) \; \delta_1 u(t) - \delta_1^T y(t) \; \delta_2 u(t)) \; dt = 0$$

For the generalization of Proposition 5.7 we proceed as follows. With the aid of Lemma 5.2 it again follows that there are a great may non-zero admissible controls $\delta u$. Let us take two admissible variations $\delta_1 u$, $\delta_2 u$ satisfying (6.23) for some $T$ and their corresponding variations $\delta_1 y$, $\delta_2 y$. Suppose that the variational condition (6.24) holds. Then, using (6.18) we obtain

$$
\begin{aligned}
0 = & \int_0^T (\delta_1^T u(t) \; ( \int_0^t W_v(t,\sigma,u) \; \delta_2 u(\sigma)d\sigma + D(t) \; \delta_2 u(t)) \\
& - \delta_2^T u(t) \; ( \int_0^t W_v(t,\sigma,u) \; \delta_1 u(\sigma)d\sigma + D(t) \; \delta_1 u(t))) \; dt \\
= & \int_0^T \int_0^t (\delta_1^T u(t) \; W_v(t,\sigma,u) \; \delta_2 u(\sigma) - \delta_2^T u(t) \; W_v(t,\sigma,u) \; \delta_1 u(\sigma))d\sigma dt \\
& + \int_0^T \delta_1^T u(t) \; (D(t) - D^T(t)) \; \delta_2 u(t)dt
\end{aligned}
$$

$(6.25)$

Using variations $\delta_1 u$, $\delta_2 u$ with support concentrated around a certain time t it follows by continuity from (6.25) that

(6.26)    $D(t) = D^T(t)$,        $t \geq 0$

Hence (6.25) reduces to

(6.27)    $0 = \int_0^T \int_0^t (\delta_1^T u(t) \, W_v(t,\sigma,u) \, \delta_2 u(\sigma) - \delta_2^T u(t) \, W_v(t,\sigma,u) \, \delta_1 u(\sigma)) \, d\sigma dt$

which is the same equation as equation (5.12) in section 5.2. The proof of propositions (5.5) and (5.7) now proceeds as before. Thus we obtain

PROPOSITION 6.5

Consider a quasi-minimal, analytic and complete system (6.1). Suppose that for any $(u,y) \in \Sigma_e^+(0)(x_0)$, all admissible variations $(\delta_1 u, \delta_1 y)$ of $(u,y)$ of compact support, $i = 1,2$, satisfy (6.24). Then, given any $T > 0$ there exists a piecewise constant control $\bar{u}$ on $[0,T]$ such that for any piecewise constant control u satisfying $u(t) = \bar{u}(t)$, $t \in [0,T]$ we have

(6.28)    $W_v(t,\sigma,u) = - W_v^T(\sigma,t,u)$,   $t,\sigma \geq T$

It therefore follows from (6.13) and (6.26) together with (6.28), that if the variational criterion (6.24) is satisfied then for a system (6.1) initialized at $\psi_{T,0}(x_0,\bar{u})$, with $\bar{u}$ as in Proposition 6.5, every variational system is self-adjoint. Combining this result with Lemma 6.4 and Theorem 6.3 we obtain the following generalization of Theorem 5.9.

THEOREM 6.6

Consider a minimal, analytic and complete system (6.1), such that the extended system is complete. Then the system is Hamiltonian if and only if for any $(u,y) \in \Sigma_e(0)(x_0)$ and admissible variations $(\delta_1 u, \delta_1 y)$, $i = 1,2$, of compact support of $(u,y)$ we have

(6.29)    $\int_0^T [\delta_1^T u(t) \delta_2 y(t) - \delta_2^T u(t) \, \delta_1 y(t)] \, dt = 0$

## 7. FINAL REMARKS AND SOME OPEN PROBLEMS

In this final chapter we shall touch upon some extensions to the theory presented in this monograph. Main emphasis is on the generalization of the mathematical tools used so far, although section 7.2 is concerned with the physical interpretation of our variational characterization.

## SECTION 7.1 Equivalence of the self-adjointness condition and Jakubczyk's algebraic condition.

In chapters 2 through 6 we have presented a characterization of Hamiltonian control systems from the behavior of variations in their input-output and state trajectories. On the other hand in chapter 1 we showed how Jakubczyk had characterized Hamiltonian systems; theorems (1.4) and (1.5), via the input-output map directly. We give below a connection between the two approaches. Our self-adjointness criterion, theorems (4.2), (4.2)' for affine systems (1.1), and theorem (6.3) for more general systems (1.2) bears the closest resemblance to the criterion stated by Jakubczyk, theorem (1.6), which is expressed in terms of the general nonlinear system (1.2). Notice that Jakubczyk's conditions comprise two parts, one of which gives conditions under which a realization of an input-output map exists at all, and the other part provides extra algebraic conditions which ensure that the (minimal) realization is Hamiltonian. One therefore expects the extra algebraic conditions to be equivalent to the conditions expressed in our theorem (6.3). However the algebraic condition given by Jakubczyk in theorem (1.6) is not particularly elegant; his characterization seems to be more suited to identifying the Hamiltonian structure expressed in equations (1.21), as in Jakubczyk's theorem (1.4). (Contrast the simplicity of equations (1.23) with the algebraic conditions given in theorem (1.6)).
For these reasons we prefer to demonstrate here that our self-adjointness criterion for affine nonlinear systems, theorems (4.2), (4.2)', characterizing Hamiltonian systems (1.15) as ones for which the variational systems are self adjoint, is equivalent to algebraic conditions on the input-output map of the system expressed in terms of the Volterra series, extended to the multivariable case and infinite Volterra series. We shall not give a direct demonstration that these conditions on the Volterra kernels; which define the input-output map of a system (1.1), imply that the system is indeed Hamiltonian, since this is essentially what Jakubczyk did.

Let us first remark that it is far from trivial to show directly that these conditions on the Volterra kernels are equivalent to our variational characterization of Hamiltonian systems stated in Theorem (5.11), namely

(7.1) $\qquad \int_0^\infty (\delta_2 y(t)^T \delta_1 u(t) - \delta_1 y(t)^T \delta_2 u(t)) dt = 0$

In Van der Schaft [V2] this is done in the case of linear systems. Systems with input-output maps defined by the equation

$$y(t) = \int_0^t \int_0^{\sigma_1} W_2(t, \sigma_1 \sigma_2) u(\sigma_1) u(\sigma_2) d\sigma_1 d\sigma_2$$

are considered in Crouch [C3]. Here the conditions amount to the following

$$W_2(t, \sigma_1, \sigma_2) + W_2(\sigma_1, t, \sigma_2) \equiv 0$$

$$W_2(t, \sigma_1, \sigma_2) + W_2(\sigma_1, \sigma_2, t) + W_2(\sigma_2, t, \sigma_1) \equiv 0,$$

and the equivalence with (7.1) is demonstrated by explicit computation, modular certain technical details which are dealt with in chapter 5. The general case must involve a heavily combinatorial type argument, such as in [La].

Consider now a system (1.1)

(7.2) $\qquad \dot{x} = g_0(x) + \sum_{i=1}^m u_i g_i(x), \qquad\qquad x \in M, \ x(0) = x_0$

$$y_i = H_i(x), \quad 1 \leq i \leq p.$$

The Volterra series expansion of the input-output map of (7.2), generalizing that given in (1.6) to multiple inputs and multiple outputs, may be written as

$$y_i(t) = W_0^{i_0}(t, x_0) + \int_0^t \sum_{i_1} W_1^{i_0 i_1}(t, \sigma_1) u_{i_1}(\sigma_1) d\sigma_1 + \ldots +$$

$$\int_0^t \int_0^{\sigma_1} \cdots \int_0^{\sigma_{k-1}} \sum_{i_1 \ldots i_k} W_k^{i_0 i_1 \ldots i_k}(t, \sigma_1 \cdots \sigma_k, x_0) u_{i_1}(\sigma_1) u_{i_2}(\sigma_2) \ldots u_{i_k}(\sigma_k) d\sigma_1 d\sigma_2 \cdots d\sigma_k$$

$$+ \ldots$$

If $(t, x) \to \Upsilon(t)(x)$ denotes the flow of $g_0$, as in chapter 1, define the time dependent vector fields $G_i(\sigma)$ as follows,

$$G_i(\sigma)(x) = \Upsilon(-\sigma)_* g_i(\Upsilon(\sigma)(x)).$$

We may now write the Volterra kernels in the form

$$W_k^{i_0 i_1 \cdots i_k}(t, \sigma_1 \cdots \sigma_k, x_0)$$

(7.3)
$$= G_{i_k}(\sigma_k)(x_0)(G_{i_{k-1}}(\sigma_{k-1})(.)(\ldots(G_{i_1}(\sigma_1)(.)(H_{i_0} \circ \gamma(t)(.))\ldots)$$

(We use the notation $G(H)$ for the Lie derivative of a function H by a vector field G.) We also generalize the definition of the bracket operation on the kernels given in (1.19) to the case of multiple inputs and outputs, by applying the bracket simultaneously to the indices, for example

$$W_k^{i_0 i_1 \cdots [i_r, i_{r+1}] \cdots i_k}(t, \sigma_1 \cdots [\sigma_r, \sigma_{r+1}] \cdots \sigma_k, x_0)$$

$$= W_k^{i_0 i_1 \cdots i_r i_{r+1} \cdots i_k}(t, \sigma_1 \cdots \sigma_r, \sigma_{r+1} \cdots \sigma_k, x_0)$$

$$-W_k^{i_0 i_1 \cdots i_{r+1} i_r \cdots i_k}(t, \sigma_1 \cdots \sigma_{r+1}, \sigma_r \cdots \sigma_k, x_0).$$

For simplicity of notation we shall often drop the time parameters, for example

$$W_k^{i_0 \cdots [i_r \cdots i_s] \cdots i_k} \quad \text{for} \quad W_k^{i_0 \cdots [i_r \cdots i_s] \cdots i_k}(t, \sigma_1 \cdots [\sigma_r \cdots \sigma_s] \cdots \sigma_k, x_0).$$

We now state the main result.

## THEOREM 7.1

Every variational system of (7.2) is self adjoint if and only if the Volterra kernels satisfy

(7.4)
$$W_k^{[i_0 \cdots i_r] i_{r+1} \cdots i_k} = (r+1) W_k^{i_0 \cdots i_r i_{r+1} \cdots i_k} \qquad \text{for } k \geq r \geq 1.$$

Before we prove this we give a simple lemma. (See also [La] for similar computations and [J7] for related results.)

## LEMMA 7.2

The conditions (7.4) hold if and only if the following conditions hold simultaneously

(7.5)
$$W_k^{[i_0 \cdots i_{r-1}] i_r i_r i_{r+1} \cdots i_k} = r W_k^{i_0 \cdots i_r i_r i_{r+1} \cdots i_k}, \qquad k \geq r > 1$$

(7.6) $\qquad W_k^{i_r[i_0 \cdots i_{r-1}]i_{r+1} \cdots i_k} = -W_k^{i_0 \cdots i_r i_{r+1} \cdots i_k}$ ,  $\qquad k \geq r \geq 1.$

Proof  For fixed k identities (7.4) are equal to the identities (7.5) except for k = r in (7.4)

$$W_k^{[i_0 \cdots i_k]} = (k+1) \, W_k^{i_0 \cdots i_k}.$$

However by definition

$$W_k^{[i_0 \cdots i_k]} = W_k^{[i_0 \cdots i_{k-1}]i_k} - W_k^{i_k[i_0 \cdots i_{k-1}]}$$

which by (7.5) with k = r and (7.6) with k = r is equal to $(r+1)W_k^{i_0 \cdots i_r}$. Thus (7.5) and (7.6) together imply (7.4).

Conversely for fixed k (7.4) for k > r ≥ 1 yields (7.5) for k ≥ r > 1. Also by definition

$$-W_k^{[i_0 \cdots i_r]i_{r+1} \cdots i_k} + W_k^{[i_0 \cdots i_{r-1}]i_r i_{r+1} \cdots i_k}$$

$$= W_k^{i_r[i_0 \cdots i_{r-1}]i_{r+1} \cdots i_k}$$

Thus (7.4), and (7.4) with r replaced by r-1 yield (7.6).  □

Proof of Theorem 7.1

We prove this result by using the input-output characterization of self-adjointness, namely that the prolongation and Hamiltonian extension of (7.1) have the same input-output map (theorem 4.2'). In the case m = p, we recall from (2.17) and (2.20) that the prolongation and Hamiltonian extension are given by the following sets of equations respectively.

$$\dot{x}_p = \dot{g}_0(x_p) + \sum_{i=1}^{m} u_i \dot{g}_i(x_p) + \sum_{j=1}^{m} u_j{}^v g_j{}^\ell(x_p)$$

(7.7) $\qquad y_i = H_i{}^\ell(x_p) \qquad\qquad 1 \leq i \leq m, \; x_p(0) = (x_0,0) \in T_{x_0} M$

$\qquad\qquad y_i{}^v = \dot{H}_i(x_p) \qquad\qquad 1 \leq i \leq m$

$$\dot{x}_e = X_{p^T g_0}(x_e) + \sum_{i=1}^{m} u_i X_{p^T g_i}(x_e) + \sum_{j=1}^{m} u_j{}^a X_{H_j{}^\ell}(x_e)$$

(7.8)

$$y_i = H_i{}^\ell(x_e) \qquad\qquad 1 \le i \le m, \; x_e(0) = (x_0, 0) \in T_{x_0}{}^* M$$

$$y_i{}^a = p^T g_i(x_e) \qquad\qquad 1 \le i \le m$$

Clearly for each input the outputs $H_i{}^\ell(x_p)$ and $H_i{}^\ell(x_e)$ coincide since they also coincide with $H_i(x)$ of (7.1) subject to the same control. It is therefore sufficient to compare the input-output maps of (7.7) and (7.8) with outputs $y_i{}^v = \dot{H}_i(x_p)$ and $y_i{}^a = p^T g_i(x_e)$ respectively. We do this by computing the Volterra kernels for each system explicitly.

We denote the Volterra kernels for the prolongation (7.7) by

(7.9)
$$W_{v,k}{}^{i_0 i_1 \cdots i_{r-1} j_r i_{r+1} \cdots}(t, \sigma_1 \cdots \sigma_{r-1}, \sigma_r, \sigma_{r+1} \cdots, (x_0, 0))$$

or simply $W_{v,k}{}^{i_0 i_1 \cdots i_{r-1} j_r i_{r+1} \cdots}$, and those of the Hamiltonian extension (7.8) as

(7.10)
$$W_{a,k}{}^{i_0 i_1 \cdots i_{r-1} j_r i_{r+1} \cdots}(t, \sigma_1 \cdots \sigma_{r-1}, \sigma_r, \sigma_{r+1} \cdots, (x_0, 0))$$

or simply $W_{a,k}{}^{i_0 i_1 \cdots i_{r-1} j_r i_{r+1} \cdots}$.

These kernels are the coefficients of the products

(7.11)
$$u_{i_1}(\sigma_1) u_{i_2}(\sigma_2) \cdots u_{i_{r-1}}(\sigma_{r-1}) u^v{}_{j_r}(\sigma_r) u_{i_{r+1}}(\sigma_{r+1}) \cdots$$

and

(7.12)
$$u_{i_1}(\sigma_1) u_{i_2}(\sigma_2) \cdots u_{i_{r-1}}(\sigma_{r-1}) u^a{}_{j_r}(\sigma_r) u_{i_{r+1}}(\sigma_{r+1}) \cdots$$

respectively. Note that we reserve the index j for the adjoint or variational controls, and the index i for normal controls and the outputs.

Computations of (7.9) and (7.10) can be done by applying the formula (7.3) to the systems (7.7) and (7.8) respectively. However there are a great many simplifications which we now describe. Let $(\sigma, x_p) \to \dot{\gamma}(\sigma)(x_p)$ denote the flow of $\dot{g}_0$, and $(\sigma, x_e) \to \gamma_H(\sigma)(x_e)$ denote the flow of $X_{p^T g_0}$. From lemma (3.2), $[\dot{g}_0, \dot{g}_1] = [\dot{g}_0, g_1]$ so

$$\dot{\gamma}(\sigma)_* \dot{g}_1(\dot{\gamma}(\sigma).) = \overline{\gamma(-\sigma)_* \dot{g}_1(\gamma(\sigma).)} = \dot{G}_1(\sigma).$$

Moreover $[\dot{g}_0, g_j{}^\ell] = [g_0, g_j]^\ell$ so

$$\dot{\gamma}(-\sigma)_* g_j{}^\ell(\dot{\gamma}(\sigma).) = (\gamma(-\sigma)_* g_j(\gamma(\sigma).))^\ell = G_j^\ell(\sigma).$$

It follows by applying (7.3) to (7.7), and using $\dot{g}_i(\dot{H}_r) = \overline{\dot{g}(H_r)}$, that if no j indices are present in (7.9)

$$W_{v,k}^{i_0 i_1 \cdots i_k} = \overbrace{W_k^{i_0 i_1 \cdots i_k}(t,\sigma_1 \ldots \sigma_k,.)}(x_0,0) \equiv 0.$$

If only one j index is present in (7.9), the use of the identities (Lemma 3.2)

$$g_j{}^\ell(\dot{H}_r) = g_j(H_r)^\ell, \quad \dot{g}_i(H_r{}^\ell) = g_i(H_r)^\ell$$

yields

(7.13)

$$W_{v,k}^{i_0 i_1 \cdots i_{r-1} j_r i_{r+1} \cdots i_k} =$$
$$= W_k^{i_0 i_1 \cdots i_{r-1} j_r i_{r+1} \cdots i_k}(t,\sigma_1 \ldots \sigma_k,.)^\ell(x_0,0)$$
$$= W_k^{i_0 i_1 \cdots i_{r-1} j_r i_{r+1} \cdots i_k}$$
$$= G_{i_k}(\sigma_k)(x_0)(\ldots(G_{i_{r+1}}(\sigma_{r+1})(.) ($$

$$G_{j_r}(\sigma_r)(.)(\ldots.(G_{i_1}(\sigma_1)(.)(H_{i_0} o\gamma(t)(.))\ldots).$$

If more than one j index is present in (7.9) the identity $g_j{}^\ell(H_r{}^\ell) = 0$ shows that the Volterra kernel is also identically zero, as in the case of no j indexes. (Note that for Volterra kernels of a system (7.2), j and i indexes play an identical role.) From lemma (3.6) $X_{p^T g_0}(p^T g_i) = p^T[g_0, g_i]$ so

$$p^T g_i \, o\gamma_H(\sigma) = p^T\gamma(-\sigma)_* g_i(\gamma(\sigma).) = p^T G_i(\sigma)(.) \quad \text{and}$$

$$\gamma_H(-\sigma)_* X_{p^T g_i}(\gamma_H(\sigma).) = X_{p^T g_i o\gamma_H(\sigma)} = X_{p^T G_i(\sigma)}.$$

It follows by applying (7.3) to (7.8) that if no j indexes are present in (7.10)

$$W_a^{i_0 i_1 \cdots i_k}$$

$$= p^T[G_{i_k}(\sigma_k)(.),[G_{i_{k-1}}(\sigma_{k-1})(.),[\ldots[G_{i_1}(\sigma_1)(), G_{i_0}(t)(.)]\ldots](x_0,0) \equiv 0$$

If only one j index is present in (7.10), the identities (Lemma 3.6)

$$X_{p^T g_0}(H_j{}^\ell) = g_0(H_j)^\ell, \quad X_{H_j}{}^\ell(p^T g_r) = -g_r(H_j)^\ell, \quad \text{and hence}$$

$$\gamma_H(-\sigma)_* X_{H_j}{}^\ell(\gamma_H(\sigma).) = X_{H_j}{}^\ell \circ \gamma_H(\sigma) = X_{H_j} \circ \gamma(\sigma)^\ell,$$

yield

$$W_{a,k}{}^{i_0 i_1 \cdots i_{r-1} j_r i_{r+1} \cdots i_k}$$

(7.14)

$$= -G_{i_k}(\sigma_k)(x_0)(\ldots G_{i_{r+1}}(\sigma_{r+1})(.)($$

$$[G_{i_{r-1}}(\sigma_{r-1})(.),[\ldots \quad \ldots,[G_{i_2}(\sigma_2)(.),$$

$$[G_{i_1}(\sigma_1)(.),G_{i_0}(t)(.)]\ldots](H_{j_r} \circ \gamma(\sigma_k).)\ldots).$$

If more than one j index is present in (7.10), we use the fact that $X_{H_j}{}^\ell(H_r{}^\ell) = 0$ to deduce that the Volterra kernel is also identically zero, as in the case of no j index.

In conclusion we see that the input-output maps of the prolongation and Hamiltonian extension are equal if and only if all the kernels in (7.13) and (7.14) coincide. If condition (7.4) is satisfied then by lemma (7.2) we know condition (7.6) is also satisfied. Using (7.6) in (7.13) via the identity

(7.15)    $$g_1(g_2(H)) - g_2(g_1(H)) = [g_1,g_2](H),$$

we deduce from (7.14) that $W_{v,k}{}^{i_0 i_1 \cdots i_r j_r i_{r+1} \cdots i_k} = W_{a,k}{}^{i_0 i_1 \cdots i_r j_r i_{r+1} \cdots i_k}$, which is sufficient to conclude that the input-output maps of (7.7) and (7.8) coincide. Conversely if the input-output maps of (7.7) and (7.8) coincide, we can equate the expressions in (7.13) and (7.14). We deduce that (7.5) holds by applying the Dynkin-Specht-Wever formula to the Lie brackets in (7.14). That (7.6) holds, may

be seen by applying (7.15) repeatedly to $W_k{}^{j_r i_0 \cdots i_{r-1} i_{r+1} \cdots i_k}$ to obtain a

formula for $W_k{}^{j_r [i_0 \cdots i_{r-1}] i_{r+1} \cdots i_k}$, which is seen to coincide with

$-W_{a,k}{}^{i_0 i_1 \cdots i_{r-1} j_r i_{r+1} \cdots i_k}$, as given in (7.14). This however is equal to

$-W_{v,k}{}^{i_0 i_1 \cdots i_{r-1} j_r i_{r+1} \cdots i_k}$ because (7.13) and (7.14) coincide. Thus since (7.5) and

(7.6) hold, by lemma (7.2), (7.4) also holds, which completes the proof. $\qquad\square$

## SECTION 7.2 A physical interpretation of the variational criterion

The basic equality which holds for (affine or general) Hamiltonian systems, and which forms the starting point for their variational characterization in chapters (5) and (6) is equation (4.36), suggestively rewritten as

(7.16)

$$\int_0^T (\delta_2^T u\, \delta_1 y - \delta_1^T u\, \delta_2 y)dt =$$

$$\omega_{x(T)}(\delta_1 x(T), \delta_2 x(T)) - \omega_{x(0)}(\delta_1 x(0), \delta_2 x(0))$$

Here $(\delta_1 u, \delta_1 y, \delta_1 x)$ and $(\delta_2 u, \delta_2 y, \delta_2 x)$ denote arbitrary (infinitesimal) variations to any solution $(\bar{u}, \bar{y}, \bar{x})$ of the Hamiltonian system under consideration.

Although this formula is very appealing, a direct physical interpretation of it is hard to find. We shall now deduce a consequence of (7.16), which has some physical meaning.

Let us consider arbitrary solutions $(\bar{u}, \bar{y}, \bar{x})$ on a time-interval (a,b) containing $[0,T]$ for which the input $\bar{u}(t)$ is identically zero. We define two variations to such a solution in the following way. First take the one-parameter variational family

(7.17) $\quad (\bar{u}(t+\epsilon), \bar{y}(t+\epsilon), \bar{x}(t+\epsilon)), \; t \in [0,T]$

By stationarity these are solutions of the Hamiltonian system for any small $\epsilon$. This results in the first (infinitesimal) variation

$$\delta_1 \bar{u}(t) := \lim_{\epsilon \to 0} \frac{u(t+\epsilon) - u(t)}{\epsilon} = \dot{\bar{u}}(t)$$

$$\delta_1 \bar{y}(t) := \lim_{\epsilon \to 0} \frac{\bar{y}(t+\epsilon) - \bar{y}(t)}{\epsilon} = \dot{\bar{y}}(t)$$

$$\delta_1 \bar{x}(t) := \lim_{\epsilon \to 0} \frac{\bar{x}(t+\epsilon) + \bar{x}(t)}{\epsilon} = \dot{\bar{x}}(t)$$

For the second variation we take any input function $F(t)$, $t \in [0,T]$, with the property that the solution $x(t)$ of the Hamiltonian system starting from initial condition $x(0) = \bar{x}(0)$ for this input function coincides at time T with the solution $\bar{x}(T)$ of the system without input ($\bar{u} = 0$). Take such a function $F(t)$ as the input variation $\delta_2 u$, and let $\delta_2 y$ and $\delta_2 x$ denote the resulting output, respectively state, variation. Now apply formula (7.16). Since $\bar{u}(t) \equiv 0$ we have $\delta_1 \bar{u} \equiv 0$.

Moreover by definition of F(t) we have $\delta_2 x(0) = \delta_2 x(T) = 0$. Hence (7.16) specializes to

$$(7.18) \qquad \int_0^T F^T(t) \, \dot{\bar{y}}(t)dt = 0$$

This equation admits the following interpretation. If u(t) can be regarded as external forces, and y(t) as displacements (as is commonly the case, see the Introduction), then $\int_0^T F^T(t) \, \dot{\bar{y}}(t)dt$ is the external work performed by the force F(t) on the system along the solution $(\bar{u} = 0, \bar{y}, \bar{x})$.

Hence (7.18) expresses the fact that <u>for any force function F(t), t $\in$ [0,T], such that the "virtual motion" $\delta_2 x(t)$ satisfies $\delta_2 x(0) = \delta_2 x(T) = 0$, the external work performed by F(t) on the time interval [0,T] is zero.</u>

The above analysis can be immediately generalized to the case that $\bar{u}$ is not identically zero, but a <u>conservative</u> force. Let us restrict ourselves to affine Hamiltonian systems $\dot{x} = X_{H_0}(x) - \sum_{j=1}^m u_j X_{H_j}(x)$, $y_j = H_j(x)$, $j = 1,\ldots,m$. Then $\bar{u}(t)$ is conservative if it is of the form

$$(7.19) \qquad \bar{u}_j(t) = \frac{\partial S}{\partial y_j}(\bar{y}(t)), \qquad\qquad j = 1,\ldots,m$$

for a certain function $S : R^m \to R$. In this case the external force $\bar{u}(t)$ can be added to the internal forces by adding to the internal Hamiltonian $H_0$ the extra "potential energy" $S(y) = S(H_1,\ldots,H_m)$. For this modified Hamiltonian system the external force is now zero, and so it follows that for any F(t), t $\in$ [0,T], as before

$$(7.20) \qquad \int_0^T (F(t) - \bar{u}(t))^T \, \dot{\bar{y}}(t)dt = 0$$

Since

$$\int_0^T \bar{u}^T(t)\dot{\bar{y}}(t)dt = \int_0^T \frac{\partial S}{\partial y}(\bar{y}(t))\dot{\bar{y}}(t)dt = S(\bar{y}(T)) - S(\bar{y}(0))$$

this yields

$$(7.21) \qquad \int_0^T F^T(t)\dot{\bar{y}}(t)dt = S(\bar{y}(T)) - S(\bar{y}(0))$$

i.e. the external work equals the increase in potential energy.

In case $\bar{u}(t)$ is arbitrary (non-conservative), the situation becomes more complicated. Consider again an input variation $\delta_2 u(t) = F(t) - \bar{u}(t)$ as before, resulting in an output variation $\delta_2 y(t)$ and state variation $\delta_2 x(t)$. Since $\delta_2 x(0) = \delta_2 x(T) = 0$ we obtain from (7.16)

$$\int_0^T (F(t) - \bar{u}(t))^T \dot{y}(t) dt = \int_0^T \dot{\bar{u}}^T(t) \delta_2 y(t) dt$$

The quantity on the right-hand side can be rewritten as

$$\int_0^T \dot{\bar{u}}^T(t) \delta_2 y(t) dt = \bar{u}^T(t) \delta_2 y(t)]_0^T - \int_0^T \bar{u}^T(t) \delta_2 \dot{y}(t) dt$$

$$= -\int_0^T \bar{u}^T(t) \delta_2 \dot{y}(t) dt$$

since $\delta_2 y(0) = \delta_2 y(T) = 0$. Hence we obtain

(7.22) $$\int_0^T F^T(t) \dot{y}(t) dt + \int_0^T \bar{u}^T(t) \delta_2 \dot{y}(t) dt = \int_0^T \bar{u}^T(t) \dot{y}(t) \, dt$$

Here the quantity on the right is the increase of internal energy of the system, while both terms on the left-hand side can be interpreted as some kind of (virtual) external work.

## SECTION 7.3 Poisson control systems and moment maps

In this final section we make some connections between the results presented above and other work on Hamiltonian mechanics/systems; indicating where these connections may either help in understanding issues not resolved by us or further the scope of the results presented.

The main mathematical construct we used here was the symplectic structure. However this may be generalized quite significantly by the use of Poisson structures. A Poisson structure on a smooth manifold M, is a bilinear form on the smooth functions, also denoted $\{.,.\}$, satisfying

(a)  $\{f, \{g,h\}\} + \{g, \{h,f\}\} + \{h, \{f,g\}\} = 0$   (Jacobi)
(b)  $\{f, gh\} = g\{f,h\} + h\{f,g\}$   (Leibniz)
(c)  $\{f, g\} = -\{g,f\}$   (anti-symmetry)

for smooth functions f, g and h on M. We call a manifold endowed with a Poisson structure a Poisson manifold, see Weinstein [We1], [We2], Van der Schaft [V7], and the book by Olver [O], for details which we sketch now.

Given a Poisson manifold we may also define the notion of a Hamiltonian vector field; $X_h$ denotes the Hamiltonian vector field defined by

$$X_h(f) = \{h,f\}$$

for every smooth function f on M, and h is the corresponding Hamiltonian function. Clearly the Poisson structure defines a linear mapping

$$B_p : T_p^*M \rightarrow T_pM, \qquad\qquad df(p) \rightarrow X_f(p)$$

whose rank determines the "rank" of the Poisson structure at $p \in M$. The rank is clearly even dimensional since $dg(B_p(df)) = dg(X_f(p)) = \{f,g\}(p)$ is skew symmetric. A Poisson structure is therefore more complex than a symplectic structure; in fact as in Weinstein [We2], a Poisson manifold is locally about $p$ a product of a symplectic manifold and a Poisson manifold of rank zero at $p$. Thus a Poisson structure with rank equal to the dimension of M everywhere determines a symplectic structure. In general the rank may vary but the distribution on M defined by the span of Hamiltonian vector fields is integrable with leaves being symplectic manifolds, on which the Poisson structure restricts to define the corresponding symplectic structure, see Olver [O].

We may now generalize the definition of a Hamiltonian system to a Poisson system, at least in the case where the control enters affinely, (see Van der Schaft [V7]), by starting with a Poisson manifold and writing the system in the usual form

$$\dot{x} = X_{H_0}(x) - \sum_{i=1}^{m} u_i X_{H_i}(x)$$

(7.23)

$$y_i = H_i(x) \qquad\qquad 1 \le i \le m$$

where $X_{H_i}$ are the Hamiltonian vector fields defined by the functions $H_i$, via the Poisson structure. The reason for introducing such a generalization lies in the fact that many physical systems may be expressed more readily as Poisson systems than Hamiltonian systems. For example mechanical systems evolving on odd dimensional manifolds would be expressed as Poisson systems first, and subsequently "reduced" to the symplectic leaves. The angular velocity equations (with or without external torques) is a good example, see Olver [O]. See also more complicated examples in Krishnaprasad ([K1]). The extension of the notion of Poisson systems to the more general class of systems discussed in chapter (6) is more subtle. In case the input and output spaces are globally distinguished, e.g. $W = Y \times U = R^p \times \Omega$, $\Omega \subset R^m$, we can define a global Poisson system on a Poisson manifold M as

(7.25)

$$\dot{x} = X_H(x,u)$$
$$\qquad\qquad\qquad\qquad u \in \Omega \subset R^m$$
$$y_j = -\frac{\partial H}{\partial u_j}(x,u), \qquad j = 1,\ldots,m$$

with $X_H(.,u)$ the Hamiltonian vectorfields defined by the functions $H(.,u)$ via the Poisson structure. In general the definition requires a development similar to that in Sniatycki and Tulczyjew [Sn], relating (locally) Hamiltonian vector fields and a suitable form of Lagrangian submanifolds of (tangent bundles to) Poisson manifolds. The idea of prolonging the Poisson structure of a manifold to its tangent bundle is dealt with in Hermann [H2]. Whereas an infinitesimal symplectic automorphism is necessarily a locally Hamiltonian vector field on a symplectic manifold, an infinitesimal automorphism of the Poisson structure on a Poisson manifold is not necessarily a locally Hamiltonian vector field, see Weinstein [We3]. This combined with the variable rank of a Poisson structure will probably make the required generalization of the affine Hamiltonian system (7.1) a more delicate matter than in the symplectic case (see [A1]).

A peculiar feature of Poisson control systems is that in case the rank of the Poisson structure in a point p, i.e. the rank of $B_p$, is less than the dimension of the manifold (or equivalently, if the Poisson structure does not define a symplectic structure), then the system cannot be accessible. This follows from the fact that the dimension at a point p of the distribution spanned by all Hamiltonian vectorfields equals the rank of $B_p$. (In fact, the maximal integral manifold through p of this distribution is the maximal symplectic leaf through p as alluded to before. ) Since the accessiblility distribution trivially is contained in this distribution, its dimension is always less than the rank of the Poisson structure. This implies, inparticular, that we may always restrict a Poisson control system with initial state to the maximal symplectic leaf through this point to obtain a Hamiltonian control system on this symplectic leaf (see [V7]).

A Poisson structure formulation of the realization problem may also provide a better setting in which to deal with another aspect of our work. Our main result for affine systems, characterizing Hamiltonian systems from their input-output data, theorem (5.11), relates to initialized systems. In a practical sense this is not particularly useful, since a physical observation of a system probably consists of input-output records from many initial states. The realization theory for non initialized systems, where one is given input-output records from some (or all) possible initial states has not been worked out; indeed very little literature at all exists on the subject. The main difficulty arises in considering input-output records corresponding to systems which are not orbit-minimal (or accessible in the analytic case), since then one has to compose input-output maps arising from initial states in different orbits. The extent to which these input-output maps may be viewed as arising from different systems, or subsystems of the same system presents a significant problem. Only when the original system yielding the input-output data is orbit minimal does this problem disappear. Since, as already mentioned, many physical systems may be modelled more naturally as Poisson systems, which can never be orbit minimal in case the rank of the Poisson structure is less than the dimension of the state space, it seems appropriate to consider the realization theory for non initialized systems in this case. This discussion is even more relevant to our results, corollary 5.15 and theorem 5.17, dealing with input-output records on $(-\infty,\infty)$ rather than $[0,\infty)$, since there we were already forced to consider non initialized systems. Our results could only be developed for orbit minimal systems.

The problems above apply equally to the general systems considered in chapter 6. However in this case, even more fundamental problems were encountered for even initialized systems. In particular we would like to deal with systems where the input and output spaces are only locally distinguished. The problem is, as pointed out in chapter (6), that the realization theory for these systems is not fully developed either. A resolution of this problem would yield a more general statement of our results, as discussed in Van der Schaft [V5], where the most general version of his conjecture is stated. We repeat it here modified in light of our foregoing observations. We begin with a nonlinear system $\Sigma$ as in definition (6.1), with external space W, and state space M. Assume W is a symplectic manifold with symplectic form $\omega_e$. We use the notation $\delta w(t)$ to denote a variation of an external trajectory $w(t)$ arising from the system, and consider trajectories on R for simplicity.

CONJECTURE

The complete minimal and analytic system $\Sigma$, such that the extended system is complete, is Hamiltonian if and only if for every external trajectory w

$$\int_{-\infty}^{\infty} \omega^e(\delta_1 w(t), \delta_2 w(t))dt = 0$$

for all admissible variations $\delta_1 w$ and $\delta_2 w$ of w, with compact support.  □

Of course we may also wish to generalize the situation to include an external space W which is a Poisson manifold, and deduce the existence of a Poisson system; but an exact formulation of this problem is yet to be made.

A particular Poisson structure exists on the dual space of any finite dimensional Lie algebra G, and is usually refered to as the Lie-Poisson structure, see for example Olver [O]. Let $c_{ij}^r$ $i,j,r = 1,\ldots,k$ be the structure constants of G relative to a basis $v_1,\ldots,v_k$. Let $G^*$ denote the dual space of G and $w_1,\ldots,w_k$ a dual basis. Define the Lie-Poisson bracket between two functions $F,H:G^* \rightarrow R$ by

$$(7.26) \qquad \{F,H\} \ (x) = x([DF(x), \ DH(x)]) \qquad x \in \mathbf{G}^*$$

where $DF(x)$, $DH(x) \in (\mathbf{G}^*)^* = \mathbf{G}$, and $[.,.]$ is the Lie bracket on $\mathbf{G}$. Here $DF(x)$ is defined by the Frechet derivative of $F$

$$DF(x)(y) = \lim_{h \to 0} \frac{F(x+hy) + F(x)}{h} \qquad x,y \in \mathbf{G}^*$$

With respect to the bases $w_i$, $v_i$ the definition (7.26) reduces to

$$\{F,H\} \ (x) = \sum_{i,j,r=1}^{k} c_{ij}^{r} \frac{\partial F}{\partial x_i}(x) \ \frac{\partial H}{\partial x_j}(x) x_r$$

where $x = \sum_{r=1}^{k} x_r w_r$ , $DF(x) = \sum_{i=1}^{k} \frac{\partial F}{\partial x_i}(x_i) v_i$ and $[v_i,v_j] = \sum_{r=1}^{k} c_{ij}^{r} v_r$.

Note that the space of linear functionals on $\mathbf{G}^*$ becomes a Lie algebra under the Lie-Poisson bracket (7.26) which is of course isomorphic to $\mathbf{G}$. This example of a Poisson structure is important since it arises as we now briefly show in the theory of Kostant-Kirillov-Souriau (see for example Wallach [Wa]). This development is also important because it leads to the generalization introduced by Jakubczyk [J5], mentioned in chapter (1), to the case of infinite dimensional Lie algebras, and his general realization theory for Hamiltonian systems.

Consider a Lie group G acting as a $C^\infty$ Lie transformation group on a smooth manifold M, $\phi: G \times M \to M$, $\phi(g,m) = g.m$. If M is also a symplectic manifold with symplectic form $\omega$, we obtain a Poisson bracket on M denoted $\{.,.\}$ as in chapter (1). We then have a linear map $\tau: C^\infty(M) \to \mathbf{X}^\infty(M)$, $\tau(\alpha) = X_\alpha$, where $\alpha$ is a $C^\infty$ function on M and $X_\alpha$ is the corresponding Hamiltonian vectorfield. $\tau$ is a Lie algebra homomorphism when $C^\infty(M)$ is viewed as a Lie algebra under Poisson bracket and $\mathbf{X}^\infty(M)$ is viewed as a Lie algebra under Lie bracket of vector fields. $\phi$ is called a symplectic action if each mapping $m \longmapsto g.m$ is a symplectic diffeomorphism of M, $g \in G$. In this case to each $v \in \mathbf{G}$, the Lie algebra of G, we can associate a locally Hamiltonian vector field $X^v$ in the usual way

$$X^v(m) = \frac{d}{dt}\Big|_{t=0} \exp tv.m$$

We obtain a Lie algebra homomorphism $\Gamma: \mathbf{G} \to \mathbf{X}^\infty(M)$, $v \to X^v$, whose image we denote by the Lie algebra **L**. We say that $\phi$ is a Poisson action if there exists a subalgebra **F** of $C^\infty(M)$ under Poisson bracket and a Lie algebra homomorphism $\lambda$ which makes the following diagram commutative

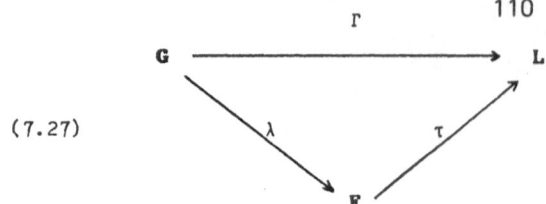

(7.27)

In case $\lambda$ existed only as a linear mapping then each $X^v \in$ is a globally Hamiltonian vector field $\Gamma(v) = X^v = X_{\lambda(v)}$. In case of a Poisson action we have also

(7.28) $\qquad \lambda([v_1,v_2]) = \{\lambda(v_1),\lambda(v_2)\} \qquad v_1, v_2 \in G.$

Let $G^*$ denote the dual of $G$ as before, then given a Poisson action $\phi$ as above we define the moment map $\mu: M \to G^*$ by setting

(7.29) $\qquad \mu(x)v = \lambda(v)(x)$

for $x \in M$, $v \in G$.

$G$ also has a natural (coadjoint) action on $G^*$, $\hat{\phi} : G \times G^* \to G^*$ defined by the following when $\ell \in G^*$ , $g \in G$ and $v \in G$,

(7.30) $\qquad \hat{\phi}(g,\ell)(v) = \ell(\text{Ad } g^{-1}(v)) \stackrel{\triangle}{=} \text{Ad}^* g(\ell)(v),$

where Ad $g : G \to G$ is the adjoint action of $G$ on $G$, defined on generators of a connected $G$ by

$\qquad\qquad$ Ad exp $v$ = exp ad $v$

and ad $v(w) = [v,w]$ is just the Lie bracket on $G$. In fact the moment map is equivariant for the actions $\phi$ and $\hat{\phi}$ of $G$ on $M$ and $G^*$ respectively; that is

(7.31) $\qquad \mu(\phi(g,m)) = \hat{\phi}(g,\mu(m)).$

Moreover since $G$ is a Lie transformation group of $M$, the orbit of $m \in M$ under $G$, denoted $M_m$, is a manifold and the orbit of $\mu(m) = p$ under $G$ in $G^*$ is also a manifold, denoted $G_p^*$.

Clearly $\mu$ maps $M_m$ onto $G_p^*$. The famous theory of Kostant-Kirillov-Souriau shows that $G_p^*$ is in fact a symplectic manifold and $\mu : M_m \to G_p^*$ is a covering map. The symplectic form on $G_p^*$ is given at $x \in G_p^*$ by

(7.32)     $\Omega_x(\hat{X}^v, \hat{X}^w) = x([v,w])$

where $\hat{X}^v(x) = \frac{d}{dt}\hat{\phi}(\exp tv, x)\big|_{t=0}$ is the vector field on $G_p^*$ induced by the action of $G$ on it. Indeed $\hat{X}^v$ is a Hamiltonian vector field on $G_p^*$ with Hamiltonian function $H_v$ defined by the linear functional

(7.33)     $H_v(\ell) = \ell(v)$

for $v \in G$ and $\ell \in G^*$.

Now the Lie-Poisson structure on $G^*$ defined in equation (7.26) yields a partition of $G^*$ into symplectic leaves, which are the orbits of the corresponding Hamiltonian vector fields determined from

$$X_H(F)(x) = \{F,H\}(x) = x([DF(x), DH(x)]).$$

In case $H = H_v$ is a linear functional on $G^*$ as in (7.33), we obtain $DH_v(x) = v$. Now

$$\hat{X}^v(F)(x) = \frac{d}{dt}F(\hat{\phi}(\exp tv, x))\big|_{t=0} = \frac{d}{dt}\hat{\phi}(\exp tv, x)(DF(x))\big|_{t=0} =$$

$$= \frac{d}{dt}x(Ad\exp{-tv}, DF(x))\big|_{t=0} = x([DF(x), v]) = X_{H_v}(F)(x).$$

Thus the Hamiltonian vector fields $\hat{X}^v$ and $X_{H_v}$ coincide, and we deduce that the symplectic leaves of the Poisson structure are just the co-adjoint orbits.

Assume now that we are given a general Hamiltonian control system, on a symplectic manifold $(M, \omega)$

(7.34)     $\dot{x} = X_H(x,u)$ ,          $x \in (M, \omega)$ , $u \in \Omega \subset R^m$, $x(0) = x_0$

with outputs either

(7.35)     $y_j = -\frac{\partial H}{\partial u_j}(x,u)$          $j = 1, \ldots, m$ ,

or

(7.36)     $y = H(x,u)$

The observation space $H_{ext}$ of (7.34) with outputs (7.35) is constructed as in chapter (6); namely the smallest Lie algebra of functions on $M$ under Poisson bracket, which contains $\{\frac{\partial H}{\partial u_j}(.,u), \ j = 1,\ldots,m, \ u \in \Omega\}$ and is closed under Poisson bracket with all functions $\{H(.,u); \ u \in \Omega\}$ and differentiations by $\frac{\partial}{\partial u_j}$, $j = 1,\ldots,m$. For simplicity we shall denote $H_{ext}$ here as $H$. Analogously, the observation space $\bar{H}$ of (7.34) with outputs (7.36) is the smallest Lie Algebra that contains $\{H(.,u); \ u \in \Omega\}$ and is closed under Poisson bracket with all functions $\{H(.,u); \ u \in \Omega\}$ and differentiations by $\frac{\partial}{\partial u_j}$, $j = 1,\ldots,m$. As a matter of fact we have the following little lemma relating $H$ and $\bar{H}$ (see for a proof [G2, Proposition 2.1])

## LEMMA

Let $q : R^m \rightarrow R^n$ be a smooth map. Then the following linear spaces are equal

$$V = \text{span } \{\frac{\partial}{\partial u_j} q(u), \ j = 1,\ldots,m, \ u \in R^m\}$$

$$V' = \text{span } \{q(u') - q(u''), \ u', \ u'' \in R^m\} \qquad \square$$

Inparticular it follows that $H$ is a subalgebra of $\bar{H}$ and that $\bar{H}$ is also given as the smallest Poisson Lie algebra containing $\{H(.,u); \ u \in \Omega\}$. Now let us assume that $\bar{H}$ (and hence $H$) is finite-dimensional, say $\bar{H} = \text{span } \{\phi_1,\ldots,\phi_N\}$, with $\phi_1$ functions on M. We can also view $\phi_1,\ldots,\phi_N$ as linear coordinate functions on the dual Lie algebra $\bar{H}^*$, which is, as we discussed above a Poisson manifold. Calculating the time-evolution of $\phi_1$ along (7.34) we obtain

$$(7.37) \qquad \frac{d\phi_i}{dt} = \{H(x,u),\phi_i\} = \sum_{k=1}^{N} a_{ik}(u)\phi_k, \quad i = 1,\ldots,N$$

for some functions $a_{ik}(u)$, since $\{H(x,u),\phi_i\}$ is contained in $\bar{H}$. Furthermore because $\frac{\partial H}{\partial u_j}(x,u) \in \bar{H}$ it follows that the outputs are of the form

$$(7.38) \qquad y_j = \sum_{k=1}^{N} c_{jk}(u)\phi_k \qquad \qquad j = 1,\ldots,m$$

Equations (7.37) with (7.38) define a control system on $\bar{H}^*$. It immediately follows that the input-output map of this control system for the initial state $(\phi_1(x_0),\ldots,\phi_N(x_0))$ equals the input-output map of the Hamitonian control system (7.34) with outputs (7.36) we started with.

For general (non-Hamiltonian) affine control systems the above construction is due to Hijab [Hi] and Fliess and Kupka [F3]. In the affine case it follows that the functions $a_{ik}(u)$ are affine in u and the $c_{jk}$ are constants, so that we obtain a bilinear control system on $\bar{H}^*$. In the Hamiltonian case $\bar{H}^*$ is a Poisson manifold and the natural mapping $x \in M \longrightarrow (\phi_1(x),\ldots,\phi_N(x)) \in \bar{H}^*$ is a Poisson mapping ([We1]), which gives that (7.37) together with (7.38) defines a <u>Poisson</u> control system on $\bar{H}^*$. In fact, since $H(x,u) \in \bar{H}$ for any $u \in \Omega$, $H(x,u)$ is of the form $\sum_{k=1}^{N} h_k(u)\phi_k$, and serves as the generating function for the Poisson control system on $\bar{H}^*$. Concluding, under the assumption that $\bar{H}$ is finite-dimensional, we can immerse a Hamiltonian control system (7.34) with outputs (7.35) into a Poisson control system on $\bar{H}^*$. Of course by taking

$$(7.39) \qquad y = \sum_{k=1}^{N} h_k(u)\phi_k \quad (= H(x,u))$$

to be output of (7.37) we immerse the modified Hamiltonian control system (7.34) with outputs (7.36) into the modified Poisson control system (7.37) with (7.39). Furthermore recall that a Poisson control system with initial state $z_0 \in \bar{H}^*$ may be always restricted to the symplectic leaf through $z_0$, which is exactly the co-adjoint orbit through $z_0$ in $\bar{H}^*$.

The construction sketched above was used by Van der Schaft [V7] in the case of affine Hamiltonian control systems in case $H$ is finite-dimensional, and builds upon an idea employed by Goncalves [G2] for the case that the Lie algebra (under Lie bracket) generated by the vectorfields $X_H(.,u)$, $u \in \Omega$, is isomorphic to $\bar{H}$. In this latter case there is immediately given a Poisson action of the Lie group $G$ corresponding to $L$ on $M$, and a corresponding momentum mapping $\mu : M \rightarrow L^*$ defined by

$$(7.40) \qquad \mu(x)v = F(x)$$

where $F \in \bar{H}$ is such that $X^v( = \frac{d}{dt}\exp tv \big|_{t=0})$ equals $X_F$. In coordinates this moment map is given as follows. Let $v_1,\ldots,v_N$ be a basis of $G$, and $w_1,\ldots,w_k$ a basis of $G^*$. If $v = \sum_{i=1}^{N} v_i a_i \in G$ then $\lambda(v) \in \bar{H}$ is given by $\sum_{i=1}^{N} a_i p_i$ where $p_i = \lambda(v_i) \in \bar{H}$, $a_i \in R$. Thus

$$(7.41) \qquad \mu(x) = \sum_{i=1}^{N} w_i p_i(x)$$

Hence we obtain the commutative diagram (7.27) with $G$ a Lie algebra isomorphic to $L$ and $H$ given by $\bar{H}$. However, in general $L$ is only isomorphic to $\bar{H}$ <u>modulo</u> the constant functions (see chapters (3) and (6)), and so we do not immediately obtain a Poisson action on $M$ and corresponding momentum mapping. This problem was completely resolved by Jakubczyk in the infinite dimensional generalization discussed below.

The interest in the idea of immersing Hamiltonian control systems into Poisson control systems lies partly in the fact that the existence, uniqueness and structure of minimal realizations of Hamiltonian systems may be attacked this way, see Van der Schaft [V7], Goncalves [G1], Jakubczyk [J5]. Also it may enable us to study realization theory for non initialized Hamiltonian systems, since the partition of the state manifold M into orbits $M_m$, is transfered into the partition of $G^*$ into the coadjoint orbits $G^*_{\mu(m)}$. This then is seen to relate to the more general view of considering Poisson systems since, as we demonstrated, the partition of $G^*$ into co-adjoint orbits is precisely the partition into symplectic leaves of the Lie Poisson structure on $G^*$. Note that there are other more complicated Poisson structures, and inparticular some arising from "infinite dimensional" systems, as in Krishnaprasad and Marsden [K2].

The geometry of the co-adjoint orbits has been studied very intensively in special cases. We mention here the work by Auslander and Kostant [Au] in the case where G is solvable, and of course that of Kirillov [Ki] in the nilpotent case. This work in turn provides details of the control system structure; see Crouch and Irving [C2] where details of Hamiltonian realizations of finite Volterra series are discussed, and Bloch [B1] for an "infinite dimensional" application.

Finally we indicate how Jakubczyk [J5] manages to extend the Konstant-Kirillov-Souriau theory to "infinite dimensional" groups G, in order to deal with the corresponding realization theory of systems (7.34), (7.36), when the Lie algebra $\bar{H}$ is infinite dimensional. Recall the situation described in chapter (1), in which the control set $\Omega$ is thought of as a set of non commuting variables and $\Omega^*$ is the set of words $x_1 x_2 \ldots x_k$, $x_j \in \Omega$. Assume we are given a symplectic manifold M, (or indeed a Poisson manifold) and a set of smooth functions $\{h_\alpha; \alpha \in \Omega\}$ on M then we may form the Poisson Lie algebra $\bar{H}$ that they generate, along with the corresponding Hamiltonian vector fields $X_{h_\alpha}$. We do not assume $\bar{H}$ is finite dimensional.

It follows that we cannot complete diagram (7.27) with $F = \bar{H}$ via a finite dimensional Lie algebra G. Rather we form the free algebra $A$ over $R$ generated by $\Omega$ (free Lie algebra G over $R$ generated by $\Omega$), identified with the sets of formal (Lie) polynomials in elements of $\Omega$. Let $A^*(G^*)$ denote the dual spaces identified with the algebra of formal power series (formal Lie series) in elements of $\Omega$. The duality is now described by

$$\langle \sum_w a_w w, \sum_w b_w w \rangle = \sum_w a_w b_w$$

which makes sense because the sum on the right is finite. (Here the summation is taken over $\Omega^*$ and $a_w$, $b_w$ are coefficients in $R$). It is now clear how to complete the diagram (7.27). The map $\lambda$ is defined on $\Omega$ by $\lambda(\alpha) = h_\alpha$, $\alpha \in \Omega$, and extended to

Lie polynomials in $G$ by insisting that $\lambda$ is a Lie algebra homomorphism from $G$ onto $\bar{H}$. For example $\lambda(\alpha_1\alpha_2-\alpha_2\alpha_1) = \{h_{\alpha_1}, h_{\alpha_2}\}$ for $\alpha_1, \alpha_2 \in \Omega$. This is possible since $G$ is a free Lie algebra. $\Gamma$ is defined in the obvious way as $\Gamma(w) = X_{\lambda(w)}$, $w \in G$.

The (formal) momentum mapping $\mu: M \rightarrow G^*$ is now easily defined as before, equation (7.5), by setting

$$(7.42) \qquad \langle\mu(x),w\rangle = \lambda(w)(x), \qquad\qquad x \in M, w \in G$$

where as in chapter(1), $\mu(x) = \sum_w \langle \mu(x),w \rangle w$. (Compare this expression with that for the finite dimensional case, equation (7.41).) Moreover a skew-symmetric bilinear form on $G^*$ can be defined as in equation (7.32) by setting

$$(7.43) \qquad \Omega_p(w,w') = \langle p,[w,w']\rangle \qquad\qquad w \in G, p \in G^*$$

where $[w,w'] = ww'-w'w$.

Jakubczyk now shows that the formal momentum $p = \mu(x)$ defines the input-output map of an analytic system (7.34), (7.36) if and only if the rank of $\Omega_p$ is finite and $\langle\mu(x),w\rangle$ define coefficients of analytic maps as explained in section (1.2). The state space of the system is clearly the co-adjoint orbit of $p$ under $G$, which is finite dimensional because of the rank condition on $\Omega_p$.

# References

[A]     R.A. Abraham, and J.E. Marsden, "Foundations of Mechanics" (Second edition), Benjamin/Cummings, Reading (1978).

[Al]    G. Alvarez-Sanchez, "Geometric Methods of Classical Mechanics applied to Control Theory", Thesis, Univ. of Calif., Berkeley (1986).

[Au]    L. Auslander and B. Kostant, "Polarizations and Unitary Representations of Solvable Lie Groups", Inventiones Math., Vol 14 (1977) pp 255-354.

[B1]    R.W. Brockett, "Control Theory and Analytical Mechanics" in Geometric Control Theory, eds. C. Martin and R. Hermann, in Vol VII of Lie Groups, History, Frontiers and Applications, Math Sci Press, Brookline (1977).

[B2]    R.W. Brockett and A. Rahimi, "Lie Algebras and Linear Differential Equations", in Ordinary Differential Equations, ed. L. Weiss, Academic Press, New York (1972).

[B3]    R.W. Brockett, Finite dimensional linear systems, Wiley, New York (1970).

[Bi]    J.M. Bismut, Mecánique Aléatoire, Lect. Not. Math., 866, Springer, Berlin (1981).

[Bl]    A.M. Bloch, "Left Invariant Control Systems on Infinite Dimensional Homogeneous Spaces", Proceedings 24 th IEEE CDC, Fort Lauderdale (1985), pp 1027-1030.

[Bu1]   A.G. Butkovskii and Yu. I. Samoilenko, "Control of Quantum Systems", Automat. Rem. Contr., Vol 40 (1979), pp 485-502.

[Bu2]   A.G. Butkovskii and Yu. I. Samoilenko, "Control of Quantum Systems II", Automat. Rem. Control., Vol 40 (1979), pp 629-645.

[C1]    P.E. Crouch and M. Irving, "On Finite Volterra Series Which Admit Hamiltonian Realizations", Math. Systems Theory, Vol 17 (1984), pp 293-318.

[C2]    P.E. Crouch and M. Irving, "Dynamical Realizations of Homogeneous Hamiltonian Systems", SIAM J. Control and Optimization, Vol 24 (1986), pp 374-395.

[C3]    P.E. Crouch, "Hamiltonian Realizations of Finite Volterra Series", in "Theory and Applications of Nonlinear Control Systems", C.I. Byrnes, A. Lindquist, eds., North-Holland (1986), pp 247-259.

[C4]    P.E. Crouch, "Dynamical Realizations of Finite Volterra Series", SIAM J. Control and Optimization, Vol 19 (1981), pp 177-201.

[F1]    M. Fliess, "Realizations of Nonlinear Systems and Abstract Transitive Lie Algebras", Bull. Amer.Math. Soc. (N.S.), Vol 2 (1980), pp 444-446.

[F2]    M. Fliess, "Réalisation Locale des Systèmes Non Linéaires, Algèbres de Lie Filtrées Transitives, et Séries Génératrices Non Commutatives", Inv. Math, Vol 71 (1983), pp 521-533.

[F3]    M. Flies and I. Kupka, "A Finiteness Criterion for Nonlinear Input-Output Differential Systems", SIAM J. Contr. and Opt., Vol 21 (1983), pp 721-728.

[G1]    J.B. Gonçalves, "Realization Theory For Hamiltonian Systems", SIAM J. Control and Optimization, Vol 24 (1986).

[G2]    J.B. Goncalves, "Nonlinear Observability and Duality", Syst. Contr. Letters, Vol 4 (1984), pp 91-101.

[Ga]    J.P. Gauthier and G. Bornard, "Existence and Uniqueness of Minimal Realizations in the $C^\infty$ case", Syst. Contr. Letters, Vol 1 (1982), pp 395-398.

[H]    R. Hermann and A.J. Krener, "Nonlinear Controllability and Observability" IEEE Trans. on Automatic Control, Vol AC-22 (1977), pp 728-740.

[Hi]    O.B. Hijab, "Minimum Energy Estimation", Ph. D. Thesis, Berkeley, (1980).

[J1]    B. Jakubczyk, "Hamiltonian Realizations of Nonlinear Systems" in "Theory and Applications of Nonlinear Control Systems", C.I. Byrnes, A. Lindquist, eds., North-Holland (1986), pp 261-271.

[J2]    B. Jakubczyk, "Existence and Uniqueness of Realizations of Nonlinear Systems", SIAM J. Control and Optimization, Vol 18 (1980), pp 455-471.

[J3]    B. Jakubczyk, "Local Realizations of Nonlinear Causal Operators", SIAM J. Control and Optimization, Vol 24 (1986), pp 230-242.

[J4]    B. Jakubczyk, "Construction of Formal and Analytic Realizations of Nonlinear Systems" in "Feedback Control of Linear and Nonlinear Systems", D. Hinrichsen and A. Isidori, eds., Springer, New York (1982), pp 147-156.

[J5]    B. Jakubczyk, "Existence of Hamiltonian realizations of nonlinear causal operators", Bull. Pol. Acad. Sci., Ser. Math., to appear (1986).

[J6]    B. Jakubczyk, "Realizations of nonlinear systems: Three approaches", in "Proceedings Conference on the Algebraic and Geometric Methods in Nonlinear Control Theory", Paris (1985), M. Fliess, M. Hazawinkel, eds. Reidel, Dordrecht (1987), pp 3-31.

[J7]    B. Jakubczyk, "Existence of Hamiltonian structure of nonlinear systems", Institute of Mathematics, Polish Academy of Sciences, Warsaw (1986).

[J8]    B. Jakubczyk, "Poisson structures and relations on vector fields and their Hamiltonians", Bull. Pol. Acad. Sci., Ser. Math., to appear (1986).

[K1]    P.S. Krishnaprasad, "Lie-Poisson Structures, Dual-Spin Spacecraft and Asymptotic Stability", Nonlinear Analysis, Theoretical Methods and Applications, Vol 7 (1984), pp 1011-1035.

[K2]    P.S. Krishnaprasad and J. Marsden, "Hamiltonian Structures and Stability for Rigid Bodies with Flexible Attachments", to appear Arch. Rat. Mech. (1986)

[Ki]    A.A. Kirillov, "Unitary Representations of Nilpotent Lie Groups", Russian Math. Surveys, Vol 17 (1982), pp 53-104.

[L]     C.M. Lesiak and A.J. Krener, "The Existence and Uniqueness of Volterra Series for Nonlinear Systems", IEEE Trans Automatic Control, Vol AC-23 (1978), pp 1090-1095.

[La]    F. Lamnabhi-Lagarrigue and P.E. Crouch, "Algebraic and Multiple Integral Identities", Arizona State University, Dept. of Electrical and Computer Engineering, Tempe (1985).

[M]    R. Marino, "Hamiltonian Techniques in Control of Robot Arms and Power Systems", in "Theory and Applications of Nonlinear Control Systems", C.I. Byrnes, A. Lindquist, eds., North-Holland (1986), pp 65-74.

[Mo]    P.M. Morse and H. Feshbach, "Methods of Theoretical Physics, Vol 1, McGraw-Hill (1953).

[O]    P.J. Olver, "Applications of Lie Groups to Differential Equations", Graduate Texts in Mathematics , Vol 107, Springer Verlag, New York (1986).

[P]    R.S. Palais, "A global Formulation of the Lie Theory of Transitive Groups", Memoires of the A.M.S., No 22, (1957).

[R]    R. Ree, "Lie Elements and an Algebra Associated with Shuffles", Ann. of Math., Vol 68 (1958), pp 211-220.

[S1]    H.J. Sussmann, "Existence and Uniqueness of Minimal Realizations of Nonlinear Systems", Math. Systems Theory, Vol 10 (1977), pp 263-284.

[S2]    H.J. Sussmann and V. Jurdjevic, "Controllability of nonlinear systems", J. Differential Equations, Vol 12 (1972), pp 95-116.

[Sa1]    R.M. Santilli, "Foundations of Theoretical Mechanics I", Springer, New-York (1978).

[Sa2]    R.M. Santilli, "Foundations of Theoretical Mechanics II", Springer New-York (1983).

[Sar]    W. Sarlet, "The Helmholtz conditions revisited. A new approach to the inverse problem of Lagrangian dynamics", Journal Physics A: Math.Gen., Vol 15 (1982), pp 1503-1517.

[Sn]    J. Sniatycki and W.M. Tulczyjew, "Generating Forms of Lagrangian Submanifolds", Indiana Univ. Math. J. Vol 22 (1972), pp 267-275.

[St]    J. Steigenberger, "On Birkhoffian Representation of Control Systems",
        Wiss. Z. TH Ilmanau, 33 (1987), Heft 1.


[T]     F. Takens, "Variational and Conservative Systems", Report ZW 7603, Univ.
        of Groningen (1976).


[Ta]    T.J. Tarn, G. Huang and J.W. Clark, "Modelling of Quantum Mechanical
        Control Systems", Math. Modelling, Vol 1 (1980), pp 109-121.


[Tu]    W.M. Tulczyjew, "Hamiltonian systems, Lagrangian systems and the Legendre
        transformation", Symp. Math. Vol 14 (1974), pp 247-258.


[V1]    A.J. van der Schaft, "System theoretic descriptions of physical systems",
        Doct. Dissertation University of Groningen, 1983, also CWI Tract No. 3,
        CWI, Amsterdam (1984).


[V2]    A.J. van der Schaft, "Hamiltonian dynamics with external forces and
        observations", Mathematical Systems Theory, Vol 15 (1982), pp 145-168.


[V3]    A.J. van der Schaft, "Symmetries, conservation laws and time-
        reversibility for Hamiltonian systems with external forces", Journal of
        Mathematical Physics, Vol 24 (1983), pp 2095-2101.


[V4]    A.J. van der Schaft, "Controllability and observability for affine
        nonlinear Hamiltonian systems", IEEE Trans. Automatic Control, AC-27
        (1982), pp 490-492.


[V5]    A.J. van der Schaft, "System theoretic properties of Hamiltonian systems
        with external forces", Proceedings Dynamical Systems and Microphysics
        (Control Theory and Mechanics) eds. A. Blaquière, G. Leitmann, Academic,
        New York (1984), pp 379-400.


[V6]    A.J. van der Schaft, "Observability and controllability for smooth
        nonlinear systems", SIAM J. Control and Optimization Vol 20 (1982),
        pp 338-354.


[V7]    A.J. van der Schaft, "Hamiltonian and Quantum Mechanical Control
        Systems", Udine (1985), to appear in Proc. 4th Int. Seminar on Math.
        Theory of Dynamical Systems and Microphysics (Information, Complexity,
        and Control in Quantum Physics), Udine 1985, eds. A. Blaquière, S. Diner,
        G. Lochak, Academic, New York (1987).

[V8]    A.J. van der Schaft, "Stabilization of Hamiltonian Systems", Nonl. An.
        Th. Meth. Appl., Vol 10 (1986), pp 1021-1035.

[V9]    A.J. van der Schaft, "Controlled Invariance for Hamiltonian Systems",
        Math. Systems Theory, Vol 18 (1985), pp 257-291.

[V10]   A.J. van der Schaft, "On Feedback Control of Hamiltonian Systems" in
        "Theory and Applications of Nonlinear Control Systems", C.I. Byrnes, A.
        Lindquist, eds., North-Holland (1986), pp 273-290.

[W1]    J.C. Willems, "System theoretic models for the analysis of physical
        systems", Ricerche di Automatica, Vol 10 (1979), pp 71-106.

[W2]    J.C. Willems and A.J. van der Schaft, "Modeling of Dynamical Systems
        using External and Internal Variables with Applications to Hamiltonian
        Systems", in Dynamical Systems and Microphysics (Geometry and Mechanics),
        Eds. A. Avez, A. Blaquière, A. Marzollo, Academic Press, New York (1982),
        pp 233-264.

[W3]    J.C. Willems, "Dissipative Dynamical Systems", Part I & Part II, Arch.
        Rat. Mech. Anal., Vol 45 (1972), pp 321-392.

[Wa]    N.R. Wallach, "Symplectic Geometry and Fourier Analysis", Math Sci Press,
        Brookline Mass, (1977).

[We1]   A. Weinstein, "Symplectic Manifolds and their Lagrangian Submanifolds",
        Advances in Math., Vol 6 (1971), pp 329-343.

[We2]   A. Weinstein, "Lagrangian Submanifolds and Hamiltonian Systems", Amer.
        Math. Soc. Vol 98 (1973), pp 377-410.

[We3]   A. Weinstein, "The Local Structure of Poisson Manifolds", J. Diff.
        Geometry, Vol 18 (1983), pp 523-557.

[Y]     K. Yano and S. Ishihara, "Tangent and cotangent bundles", Dekker,
        New York (1973).

# Lecture Notes in Control and Information Sciences

Edited by M. Thoma and A. Wyner

# Lecture Notes in Control and Information Sciences

Edited by M. Thoma and A. Wyner

# Lecture Notes in Control and Information Sciences

Edited by M. Thoma and A. Wyner